Dry Stone Retaining Structures

Discrete Granular Mechanics Set

coordinated by
Félix Darve

Dry Stone Retaining Structures

DEM Modeling

Eric Vincens
Jean-Patrick Plassiard
Jean-Jacques Fry

ELSEVIER

First published 2016 in Great Britain and the United States by ISTE Press Ltd and Elsevier Ltd

ISTE Press Ltd
27-37 St George's Road
London SW19 4EU
UK

www.iste.co.uk

Elsevier Ltd
The Boulevard, Langford Lane
Kidlington, Oxford, OX5 1GB
UK

www.elsevier.com

For information on all our publications visit our website at http://store.elsevier.com/

British Library Cataloguing-in-Publication Data
A CIP record for this book is available from the British Library
Library of Congress Cataloging in Publication Data
A catalog record for this book is available from the Library of Congress
ISBN 978-1-78548-080-5

Printed and bound in the UK and US

Contents

Foreword

Molecular dynamics is recognized as a powerful method in modern computational physics. This method is essentially based on a factual observation: the apparent strong complexity and extreme variety of natural phenomena are not due to the intrinsic complexity of the element laws but due to the very large number of basic elements in interaction through, in fact, simple laws. This is particularly true for granular materials in which a single intergranular friction coefficient between rigid grains is enough to simulate, at a macroscopic scale, the very intricate behavior of sand with a Mohr–Coulomb plasticity criterion, a dilatant behavior under shearing, non-associated plastic strains, etc. and, at fine scale, an incrementally nonlinear constitutive relation. Passing in a natural way from the grain scale to the sample scale, the discrete element method (DEM) is precisely able to bridge the gap between micro- and macro-scales in a very realistic way, as is verified in many mechanics labs today.

Thus, DEM is in an impetuous development in geomechanics and in the other scientific and technical fields related to grain manipulation. Here lies the basic reason for this new set of books on "Discrete Granular Mechanics", in which not only numerical questions are considered but also experimental, theoretical and analytical aspects in relation to the discrete nature of granular media. Indeed, from an experimental point of view, computational

---

Foreword written by Félix DARVE.

tomography – for example – is giving rise to the description of all the translations and rotations of a few thousand grains inside a given sample and to the identification of the formation of mesostructures such as force chains and force loops. At a theoretical level, DEM is also confirming, informing or at least providing some theoretical clues as to failure modes, the expression of stresses inside a partially saturated medium and the mechanisms involved in granular avalanches. Effectively, this set of books plan to cover all the experimental, theoretical and numerical approaches related to discrete granular mechanics.

The observations show undoubtedly that granular materials have a double nature, that is continuous and discrete. Indeed, roughly speaking, these media respect the rules of continuity of matter at a macroscopic scale, whereas they are essentially discrete at the granular microscopic scale. However, it appears that, even at the macroscopic scale, the discrete aspect is still present. An emblematic example is constituted by the question of shear band thickness. In the framework of continuum mechanics, it is well recognized that this thickness can be obtained only by introducing a so-called "internal length" through "enriched" continua. However, this internal length is not intrinsic and constitutes a kind a constitutive relation by itself. The reason for this is probably that using a simple scale to determine the discrete nature of the medium oversimplifies reality. However, with DEM modeling, this thickness is obtained in a natural way without any *ad hoc* assumption. Another point, whose proper description was indomitable in a continuum mechanics approach, is post-failure behavior. The finite element method, which is essentially based on the inversion of a stiffness matrix, which becomes singular in a failure state, has some numerical difficulties going beyond a failure state. Here also, it appears that DEM is able to simulate fragile, ductile, localized or diffuse failure modes in a direct and realistic way – even in some extreme cases such as fragmentation rupture.

The main limitation of DEM today is probably linked to the limited number of grains or particles, which can be considered in relation to an acceptable computation time. Thus, the simulation of boundary value problems remains bounded by more or less heuristic cases. So, the current computations in labs involve at best a few hundred thousand

grains and, for specific problems, a few million. Let us note however that the parallelization of DEM codes has given rise to some computations involving 10 billion grains, thus widely opening the field of applications for the future.

This set of books will also present the recent developments occurring in micromechanics as they are applied to granular assemblies. The classical schemes consider a representative element volume. These schemes are proposing to go from the macro-strain to the displacement field by a localization operator, then the local intergranular law relates the incremental force field to this incremental displacement field, and eventually a homogenization operator deduces the macro-stress tensor from this force field. The other possibility is to pass from macro-stress to macro-strain by considering a reverse path. So, some macroscopic constitutive relations can be established that properly consider an intergranular incremental law. The greatest advantage of these micromechanical relations is probably that they only consider a few material parameters, each one with a clear physical meaning.

This set of around 20 books has been envisaged as an overview of all the promising future developments mentioned in this foreword.

Félix DARVE
November 2015

Introduction

Many systems in civil engineering involve material discontinuities or can exhibit large or localized deformations. These characteristics make the modeling of their mechanical behavior complicated at the time when quantitative results are expected. The discrete element method (DEM) can be very helpful in this respect, since it allows individual bodies that are interacting to be handled in a rather simple way.

Indeed, DEM has been intensively used to address the behavior of granular materials but has proven efficient for studying the behavior of larger civil engineering systems such as rockfill dams [DEL 06, SIL 09], protection against rock fall impact [NIC 07, PLA 10a, SAL 10], slope analysis [LOR 09, TAN 09, MOL 12, LIU 13, BON 15] and fractured rock mass behavior [MOA 08, ZHA 11, HAR 12, NOO 14].

Dry stone retaining structures, such as dry stone retaining walls (DSRWs) and rockfill dams with facing, are composed of stone blocks whose size generally ranges from a few centimeters to 50 cm. The blocks may be dumped or be hand-placed following a certain know-how and no mortar is used to bond any elements of these structures. If friction at contact between the blocks is then the main mechanical characteristic that justifies their overall stability, the strong interlocking between blocks is acknowledged to contribute to their strength. Due to the relative movements between blocks, they can bear

Introduction written by Eric VINCENS.

large deformations before failure, which differentiates these structures from other civil engineering structures involving, for example, reinforced concrete.

The construction of dry stone retaining structures does not involve sophisticated processes and it may explain why, in the case of DSRWs, their existence has been found worldwide for centuries. They have mainly been used to counter slope erosion, providing areas for agricultural culture, but they were also used to bear strategic works such as roads, fortresses and castles. More particularly, rockfill dams with timber upstream facing were built by gold diggers and pioneers to allow water storage in the USA in the 19th Century. Rockfill dams with dry masonry facing on the upstream slope and stone pitching on the downstream slope appeared in the early 20th Century in Italy and France.

Large permanent deformations were sometimes observed on dry stone retaining structures. In particular, over time, permanent linear deformations have been systematically monitored on rockfill dams, and movements between blocks locally observed. If in the former case they may be due to a weak foundation that deforms under the weight of the structure, they also come from the alteration or the damage of the rock material. The question that arises is: "Do these displacements indicate a decreasing safety factor of stability? May these pathologies be expected from such aging structures? What kind of maintenance works ought to be addressed?".

Due to the complex behavior of these structures, the lack of standards for design and repair, the owners need a scientific assessment of the current state to make a decision with regards to their repair.

This book makes a synthesis of two decades of research on DSRWs and rockfill dams with stone pitching with the aim of highlighting how DEM has contributed to a breakthrough in the understanding of their mechanical behavior. Chapter 1 deals with the behavior of slope and DSRWs and Chapter 2 with the behavior of rockfill dams with steep slopes and dry masonry pitching.

Dry Stone Retaining Walls

1.1. Introduction

For centuries, dry stone retaining walls (DSRWs) have been used either to support a natural slope, to counter slope erosion and allow agriculture, or to support civil engineering structures. In the latter case, we can cite the fortress of Sacsayhuaman (15th Century) or many castles or temples in Japan such as Mastuyama castle (17th Century) where ashlar masonry proved able to resist high-intensity earthquakes. But in this instance, the dressed stones allow the wall density to be maximized, while the perfect interlocking between blocks provided by contact planes provides an exceptional hardiness for such structures. This high-quality masonry is not the most common dry stone work, which is instead generally composed of rubble where the blocks are hardly recalibrated or dressed. These latter works are porous, with a porosity up to 20%, and few contact points between the blocks make relative movements between them rather likely. It may be a definite advantage, since they do not constitute an impermeable barrier against natural water flows. The typical geometry of DSRWs is shown in Figure 1.1. The wall height is generally between 2 and 4 m, but can rise up to 10 m high [ALE 12]. A batter on the outward wall face of between 10 and 15° is typically observed, which generally also corresponds to the inclination of the stone layers.

Chapter written by Eric VINCENS.

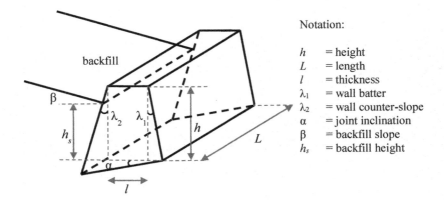

Figure 1.1. *Geometry of dry stone retaining walls [OET 14]*

Though extensively used for centuries, the development of bonded masonry or reinforced concrete structures put an end to the use of DSRWs by the early 20th Century. Nevertheless, a survey by Odent [ODE 00] showed that approximately 18% of retaining walls in France are classified as DSRWs, and O'Reilly *et al.* [ORE 99] estimated that there is around 9,000 km of such walls in Great Britain.

The consequence of this disinterest is noticeable in existing DSRWs, where little maintenance and repair has been carried out through the decades. At the time when repairs are critical, the owners are in fact helpless to make any decision due to the loss of scientific knowledge about the mechanical behavior of DSRWs. Moreover, no established standard can provide any methodology for the design, maintenance and repair of such works.

Recent decades have witnessed the emergence of new research trying to bridge the gap between the existing empirical know-how of stone masons and a more rational scientific knowledge in relation to DSRWs. In this respect, we can cite the full-scale experimental campaigns that have been carried out by two research groups, one from ENTPE, France [VIL 07, COL 10a, LE 13] and another one from the University of Bath, UK [MUN 10].

There are two kinds of DSRWs in relation to the loading they bear. Slope DSRWs just bear the loading provided by the backfill they stabilize (Figure 1.2, top wall). Highway DSRWs in addition to the backfill loading also bear concentrated loadings coming essentially from the weight of vehicles traveling on the road (Figure 1.2, bottom wall). Generally, the road is built just at the top of the wall and parallel to the wall facing.

Figure 1.2. *Two types of DSRWs: slope-retaining wall on top; highway-retaining wall at the bottom. Photo credit: Paul McCombie*

For plane slope DSRWs, which are the most frequently investigated structures (through full-scale and small-scale experiments), two modes of failure were identified: a sliding mode and a toppling mode. The latter mode of failure was found to be predominant in actual down-scaled experiments (Figures 1.3(a) and (b)) [COL 10b]. Here, failure is due to an excessive backfill pressure compared to the wall strength. In the case of plane DSRWs, a plane strain mode of failure was found which allows a two-dimensional (2D) modeling of such mechanical problems [COL 10b].

The failure typical of highway DSRWs has only recently been investigated [LE 13, OET 14, QUE 15]. In this case, failure is due to an excessive concentrated loading at the backfill surface and is characterized by a true three-dimensional (3D) state of deformations where a bulge develops before failure.

DSRW structures are discrete in nature, able to dissipate energy by friction at contact between blocks, while large deformations are generated before failure (Figure 1.3). The discrete element method (DEM) therefore seems a suitable approach to model the behavior of such a system. The modeling of two kinds of DSRWs is presented in this chapter: the case of plane slope DSRWs and the case of plane highway DSRWs.

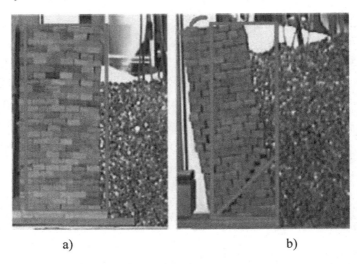

a)   b)

Figure 1.3. *Plane-strain failure of scaled down DSRW: a) sliding; and b) toppling [COL 10b]. Photo credit: Anne-Sophie Colas. For a color version of the figure, see www.iste.co.uk/vincens/drystone.zip*

1.2. Plane slope dry stone retaining walls

1.2.1. *Experimental campaigns*

The numerical work presented herein is based upon full-scale experiments performed on plane DSRWs. Following the work of Villemus *et al.* [VIL 07] who applied a hydrostatic load at the inward wall face, Colas *et al.* [COL 10a] loaded four DSRWs that were 2 m high and 2 m long with a backfill composed of rounded-elongated gravel. This granular material was dumped little by little away from the wall inducing a slope for the backfill equal to the natural angle of repose for the material. A toppling mode of failure was observed and the lengthwise view was sufficient to validate a behavior with a plane strain condition. The critical backfill height at the inward wall face inducing failure was measured. More details can be found in [COL 10a].

1.2.2. *Full DEM approach*

A slope-retaining wall is composed of three subsystems with specific mechanical properties: the wall, the backfill that may have some degree of cohesion, and the backfill–wall interface (Figure 1.4).

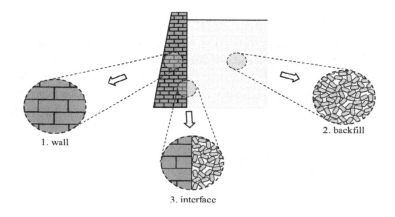

1. wall

2. backfill

3. interface

Figure 1.4. *Three subsystems for DSRWs*

We present herein the modeling of the full-scale experiments carried out by Colas *et al.* [COL 10a] where the backfill is composed of a purely frictional material. The modeling of the full-scale experiments performed by Villemus *et al.* [VIL 07] can be found elsewhere in [OET 15].

Itasca code PFC2D^{TM} (Particle Flow Code) was used to model the loading of DSRWs with a backfill. PFC2D^{TM} allows the modeling of the interaction of rigid bodies with deformable contacts using an explicit solution method to solve the dynamics equations [CUN 79]. The basic elements are rigid disks but more complex shapes for the bodies can be created by gluing various disks together. The final assembly of these glued disks can then be handled as a rigid or a deformable single body [ITA 08].

At each time step, the calculation runs in alternate between contact-force law and the law of motion. The constitutive law at microscale is at the heart of particle-based simulation. The simplest contact law that can be selected consists of a spring, where the force magnitude is calculated as follows:

$$F_n = k_n\, u_n$$

$$\Delta F_s = -k_s\, \Delta u_s$$

The normal stiffness k_n is a secant modulus, relating the total normal displacement u_n to the normal force F_n. The shear stiffness k_s is a tangent modulus relating the incremental shear displacement Δu_s to the increment of the tangential force ΔF_s. This stiffness should be sufficiently high to avoid excessive interpenetration, which may lead to an overrepresentation of the elastic behavior. At the same time, both normal and tangential stiffnesses must be small enough to reduce the critical time step. In addition to elasticity, for the modeling of purely frictional granular material, the local tangential force has a limited maximum value to satisfy a Coulomb's friction law:

$$|F_s^t = F_s^{t-1} + \Delta F_s| \leq \mu\, F_n$$

with μ denoting the local friction coefficient. When this limit is reached, sliding between the two bodies in contact can take place.

More sophisticated contact law can be envisioned to take into account more complex phenomena in granular materials, which is not considered in the present work.

1.2.2.1. Model geometry

The two subsystems, wall and backfill, are composed of bodies of different shapes. While wall blocks in the full-scale experiments roughly held parallelepiped shapes, the backfill grains were elongated flat and rounded material.

A block is recreated using a certain number of small disks which are poured within a rectangular box holding the desired dimensions. The disks are glued together forming a rigid body with deformable contacts. The inner disks of the block are erased, keeping the disks that form the outer shape of the block (Figure 1.5). The block has a natural geometrical roughness due to the disks, but the wall effects during the deposit of the disks within the box induce a further irregularity in the block profile. Disks with a diameter of between 8 and 11 mm have been used, which is a compromise between the number of particles involved and the representation of the roughness of the block created.

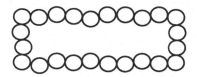

Figure 1.5. *Block model for the wall*

However, backfill grains having a given aspect ratio (AR) (ratio between the largest and the smallest grain dimension) have been created by gluing two or three particles together (some examples are given in Figure 1.6). Trying to give a physical meaning to the modeled backfill particle shape is not as simple as for a block. The AR of an actual elongated flat particle may range, for example, between 2 and 10 depending on the cutting plane considered. Then, the shape is generally calibrated in such a way that the global frictional mechanical properties of the granular material can be retrieved [DEL 06, TRA 06].

AR=1.25   AR=1.75   AR=2.5

Figure 1.6. *Grains with various aspect ratios*

A few studies have shown that grain elongation in particular affects the critical friction angle of a granular soil [ROT 92, TIN 95, MIR 02, NOU 05, NOU 10, AZE 10]. The critical state, which is the state at which a large deformation of the material can occur, is a peculiar state since, when reaching this state, the mechanical properties of the granular medium just depend on intrinsic characteristics: grading, D_{50} (size of the 50% smaller than mean size), sphericity and angularity of grains and finally the local friction angle between grains [THO 00, YAN 12]. Thereafter, the aspect ratio of the backfill grains is identified in order to obtain a critical friction angle for the modeled backfill grains that are close to the actual backfill material used in the full-scale experiments by Colas *et al.* [COL 10a]. The identification of the backfill grain AR is presented in section 1.2.2.3.2.

1.2.2.2. *Primary local mechanical properties*

The basic contact law in PFC2D^{TM} is a linear contact law where sliding can occur following Coulomb's friction law, perpendicular to the direction of the contact vector. Both normal and tangential stiffnesses must then be defined. For the sake of simplicity, a tangential stiffness has been chosen equal to normal stiffness in the present work.

Contact between individual bodies, say between wall blocks, between backfill grains or between blocks and backfill grains in the wall-backfill interface, is identified through individual instances of contact between the disks. In the case of wall blocks, this is performed through the disks that form the outer shape of the blocks (Figures 1.7(a) and (b)). Each individual contact has its own plane of contact with its own direction. A primary study showed that the block's macroroughness creates a fluctuation in contract results. It makes the identification of an average behavior for interacting bodies of different kinds, as in the case of the DSRWs, difficult or imprecise. For example, the fluctuation of the maximum mobilized friction angle throughout an interface shear test (slide of a block against a fixed block) was found to be as high as ±5°. Consequently, another contact law was chosen for block–block contacts.

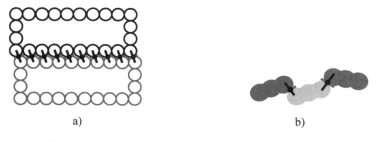

Figure 1.7. *Linear contact: a) between wall blocks and b) between backfill grains. For a color version of the figure, see www.iste.co.uk/vincens/drystone.zip*

A smooth-joint contact law [MAS 08] which basically removes the block's geometrical roughness has been chosen (Figure 1.8). This contact law resembles the linear contact law except that: (1) the contact plane direction is the same and imposed at each disk contact; and (2) sliding takes place along a given contact area A (or length in a 2D model). In 3D, the following formulations are used to calculate the contact forces for smooth-joint contacts:

$$F_n^{t} = F_n^{t-1} + k_n{'} A\, u_n$$

$$F_s^{t} = F_s^{t-1} - k_s{'} A\, \Delta u_s$$

By default, the total surface area A (or length in 2D) of a smooth joint contact is computed a using the minimum radius between two particles in contact. Due to the physical roughness, the total surface area in contact can be greater than the plane block surface area, hence A is multiplied by the ratio between the plane surface area of the block and the total surface area of contact between two blocks. A corresponding correction involving length applies in 2D. In Figure 1.8, the contact plane at each point of contact is equal to the horizontal plane, which can be considered as the average contact plane between the two blocks. Consequently, this law is similar to a homogenized contact law. For smooth-joint contacts, the macroroughness is then not taken into account, which removes the before-mentioned fluctuations.

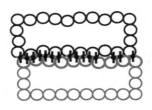

Figure 1.8. *Smooth-joint contact law between wall blocks*

1.2.2.3. *Identification of local mechanical properties*

Since there is not always a clear or direct relationship between the local behavior and the global behavior of bodies in interaction, the micromechanical properties are generally identified using a trial-and-error method. A correct set of local parameters is then supposed to correctly retrieve the global behavior of the system as observed on site.

The contact stiffnesses can be identified through shear tests in the case of block–block contacts or block–backfill grain contacts, and compression biaxial tests in the case of backfill grain–grain. Since high values for stiffnesses may lead to longer computation times, smaller values may be chosen if the overall behavior of the system is not excessively affected by this modification. For the block–block contact, a primary parametric study showed that the likelihood of the wall to fail is not modified if the smooth-joint contact stiffnesses (k_n' and k_s') are dropped to a value of 10^8 Pa/m. For the backfill grain–grain contact, the linear contact stiffnesses (k_n and k_s) were taken as being equal to 5.10^7 Pa. This latter set of parameters allows us to have an evolution of the mobilized friction angle throughout simulated biaxial tests close to the one observed throughout the actual triaxial tests made on the backfill material used in the full-scale experiments. For the wall–backfill interface, the contacts between the wall blocks and the backfill grains are linear contacts and the contact stiffnesses were also set at 5.10^7 Pa.

To identify the other local mechanical parameters, the three subsystems (the wall, the backfill and the wall–backfill interface) involved in the full-scale experiments are then successively studied and calibrated using DEM simulations.

1.2.2.3.1 Wall block contacts

In the case of block–block contact, the friction coefficient is the only local parameter that must be identified. Since for this contact, the smooth-joint law that is used is similar to a homogenized law, the local friction angle of the block–block contact ϕ_b^l is supposed to be equal to the global friction angle ϕ_b^g determined by a tilt test. This point was verified and can be found in [OET 15].

In the full-scale experiments, two different materials were used to build the DSRW: schist and limestone. Tilt tests on blocks were performed and led to block–block friction angles φ_b^g equal to 25 and 35° for schist and limestone, respectively.

1.2.2.3.2 Backfill

The grading of the fluvial material used in the full-scale experiments was between 8 and 16 mm (Figure 1.9). Since the material was gradually and carefully dumped from a dumper away from the wall, the material was expected to be in a loose state forming a slope whose angle was close to the angle of repose. This angle was measured on site and found to be close to 32°. A unit weight of 14.9 kN/m^3 for the gravel was also identified for a loose state. Then, three triaxial tests were performed in a non-conventional triaxial apparatus so that the size of the cell was greater than the representative elementary volume [COL 10a]. Three confining pressures of 20, 50 and 100 kPa were considered. Though a plateau close to 37° was observed for the mobilized friction angle at large deformations that would correspond to a critical state, this value was not retained. In fact, there was no clear confidence that no localized deformations occurred during the triaxial tests. Then, in the present work, a critical friction angle equal to the angle of repose for the backfill slope (32°) was considered. Finally, from these tests, and stating that an average representative mean stress for the backfill was equal to 35 kPa (maximum wall height of 2 m), an internal friction angle of 37.7° and a dilatancy angle of 8° for the backfill were identified for the backfill material used on site.

Figure 1.9. *Granular backfill used in the full-scale experiments by Colas et al. [COL 10a]*

Since two characteristics must be identified for the backfill, say the particle AR and the local friction angle, at least two pieces of information should be required on site. The critical friction angle (stated as being equal to the angle of repose of the backfill slope) and the internal friction angle derived from the triaxial tests were used as references. Then, the local parameters were identified and processed in two steps to retrieve these two global friction angles.

First, a trial-and-error method was used to retrieve the critical state angle. In order to save computation time, the grading of the modeled backfill material was first reduced to the interval of 10–16 mm. Two properties of the modeled granular material were likely to influence the friction angle at critical state: the grain AR and the local friction angle. Consequently, we noted that the set of local parameters was not unique and that the two parameters that must be identified could not be chosen independently.

Simulations of biaxial tests were performed on samples composed of the modeled backfill grains having a given AR and a given local friction angle between grains. The grains were deposited under gravity within a box whose size varied according to the AR of the grains involved. For example, for strongly elongated grains having an AR of 2.5, more than 2,250 grains have been used to create a sample. In this case, the grains were poured in a rigid box of 75×150 cm^2 with a drop height equal to five times the size of the maximum grain size (D_{max}). The equilibrium for the sample was computed before performing a compression test. A confinement stress of 35 kPa corresponding to the representative confining stress on site and a loading strain rate of 10^{-3}/s was chosen for the simulations so that a quasi-static condition was observed for each sample configuration [OET 14].

The results for a first set of biaxial simulations and corresponding to particles having an AR of 1.75 are given in Figure 1.10(a). The critical friction angle $\phi_{gc}{}^g$ is plotted for different local friction angles between grains $\phi_g{}^l$. Generally, the global friction angle increases with the local friction angle, nevertheless it seems that a plateau is obtained for local friction angles greater than 20°. In this case, a critical friction angle of 32° cannot be reached. Since the critical friction angle increases with the AR, further simulations were carried out using an aspect ratio of 2.5 (Figure 1.10(b)). The same features as highlighted before were obtained; nevertheless, the critical state friction angle was found to be close to 32°. We may claim that grains with an even greater AR should be considered. Nevertheless, this latter choice would definitely present drawbacks. First, a large AR generates strong local heterogeneities [NOU 03]. For the backfill–wall interface, it means fewer contact points, arching effects along the inward wall face leading the wall to be more sensitive to the backfill generation. Then, it was decided not to increase the AR of grains greater than 2.5 and the local friction angle $\varphi_g{}^l$ was selected to model the backfill granular material using a trial-and-error method. Such a high local friction may not hold a physical meaning, however, this value previously was used to model the behavior of elongated aggregates in a rockfill dam [TRA 06].

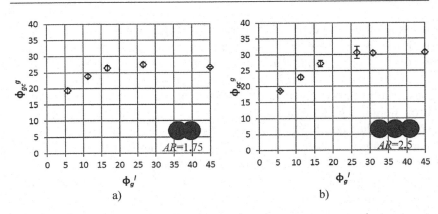

Figure 1.10. *Influence of grain-grain local friction angle ϕ_g^l on the global critical friction angle ϕ_{gc}^g for two aspect ratios: a) AR=1.75 and b) AR=2.5*

The second property for the backfill to retrieve is the internal friction angle which drives the overall shearing resistance of the backfill and then its ability to limit the pressure acting on the wall. For granular materials, this maximum shearing resistance mainly depends not only on the same properties as evoked for the critical friction angle but also on the density of the material [BOL 86, SAL 09]. A trial-and-error method is also used to identify the initial porosity of a numerical sample compatible with an internal friction angle ϕ_g^g of 37.7° by means of simulations of biaxial tests.

The deposit of grains under gravity for the backfill is supposed to induce a very loose state for the material. Nevertheless, it does not allow us to obtain the requested internal friction angle and a denser initial state must be considered. This can be achieved by gradually setting the local friction angle ϕ_g^l to a lower value, while a new equilibrium for the material is computed for each different local friction angle. When both the desired porosity and an equilibrium are reached, ϕ_g^l is reverted to 45° as previously identified. The sample is considered in static equilibrium when the ratio between the maximum unbalanced force and the maximum contact force is smaller than 10^{-4}.

For a numerical sample composed of backfill grains, a trial-and-error method is used to identify which initial porosity is able to give an internal friction angle of ϕ_g^g of 37.7°. Figure 1.11 depicts the values for the internal friction angle obtained through simulations of biaxial tests for different initial sample porosities. For each targeted porosity, three samples are created (Figure 1.11). Finally, an initial porosity n_0 of 0.22 is identified as compatible with a backfill internal friction angle ϕ_g^g of 37.7°.

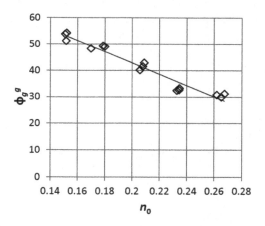

Figure 1.11. *Relation between the internal friction angle ϕ_g^g and initial porosity n_0 for a sample composed of backfill grains; AR=2.5 and $\phi_g^!=45°$*

Figure 1.12(a) gives the evolution of the stress ratio q/p (q and p are, respectively, the second invariant and the first invariant of the stress tensor) for the numerical sample made up of backfill grains and the selected parameters ($AR=2.5$, $\phi_g^!=45°$ and $n_0=0.22$). This curve was compared to the triaxial tests performed on the actual backfill material. For the actual triaxial tests, two results obtained for tests with a confining stress of 20 and 50 kPa, respectively, are given. We are reminded that the representative confining stress considered for the backfill was taken as being equal to 35 kPa.

In Figure 1.12(a), we can note that the maximum stress ratio that was obtained through numerical biaxial tests was close to the one obtained for the actual backfill material. Nevertheless, a strong softening that does not exist for the actual triaxial tests can be noticed, which calls for two remarks to be made. An initial porosity of 0.22 induces a behavior for the numerical sample that can be qualified as that of a rather dense material. As a result, the associated dilatancy was stronger than the one corresponding to the actual material and a confining stress that would be equal to 35 kPa (Figure 1.12(b)). This feature is typical of 2D samples, where dilative patterns are generally overrepresented [MIR 02]. However, the dilatancy that is observed on the actual sample was not compatible with a non-existent softening for the stress path. Consequently, the critical state that is likely to be identifiable in Figure 1.12(a) may not be the true one as previously stated.

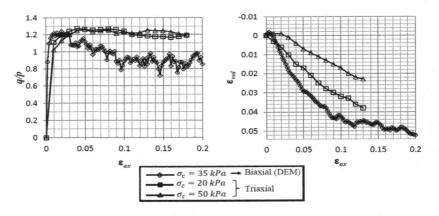

Figure 1.12. *Evolution of the stress ratio q/p and the volumetric deformations for the biaxial DEM simulation and the actual triaxial tests*

The only way to reduce the softening for the numerical sample involves proposing a new set of local parameters for the local friction angle and the AR. In particular, a stronger AR for the grains would increase the critical state angle. As mentioned before, this option is not desirable since strong local heterogeneities in terms of porosities would be generated in the system and that would be fewer contact points between the backfill and the wall.

To conclude, the identification process for the backfill leads to the following set of local parameters: an AR for the grains of 2.5, a local friction angle ϕ_g^l of 45° and a porosity n_0 of 0.22.

1.2.2.3.3. Backfill-wall interface

A soil–structure interface is a subsystem that is not easy to clearly define. First, in some cases there may be a geometrical interface which is the true boundary between the soil and a structure. This is the case for the backfill–wall interface. For the DEM modeling, it involves all the contacts between the particles from the backfill and the disks forming the outer surfaces of the wall blocks. However, the soil–structure interface can also be identified as a zone within the soil where shear stresses concentrate when the whole system is submitted to a shearing process. This definition is vague, since the influence of the structure within the soil tends to vanish gradually as the distance to the structure increases.

This zone is then identified with a device equivalent to a direct shear box [MOR 07]. From a direct shear test, the interface width was found to be within the range of $10–12D_{50}$ (D_{50} is the diameter of the 50% finer) [PRA 13].

In fact, the mechanical behavior of a soil–structure interface depends on the normalized roughness R_n of the interface (Figure 1.13) [KIS 87]. This latter property is defined as the ratio of the geometrical roughness of the structure R_{max} to the D_{50} of the soil grains. If R_n is small (typically smaller than 0.02), the interface is considered a smooth interface and displays the behavior of an elastic perfectly plastic material [FIO 02] without any dilative behavior. When the normalized roughness increases up to 0.1, some degree of dilatancy can appear which reveals a transitional behavior toward a true rough interface. When the normalized roughness is greater than 0.1, the interface exhibits an elastoplastic behavior with hardening and there may be some degree of softening toward a critical state [UES 86, FIO 02]. In this case, the interface, which is a rough interface, displays the typical behavior held by a soil, with dilatancy and density influencing of the mechanical behavior.

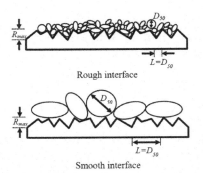

Figure 1.13. *Main properties of a rough and smooth interface [UES 86, MOR 07]*

The average normalized roughness of a backfill–wall interface was measured on site and was found to be close to 1.5, which qualifies the interface as being a (very) rough interface. No information on stresses was monitored when the full-scale experiments by Colas *et al.* [COL 10a] were carried out. It is therefore not possible to know the value of the internal friction angle of the backfill–wall interface ϕ_i^g. Nevertheless, the case of soil–DSRW interface seems very similar to that of concrete–soil interface for *in situ* concrete which also qualifies as a very rough interface. As such, following the European Standard for Geotechnical Works (EUROCODE 7), the internal friction angle of the backfill–wall interface ϕ_i^g can be chosen to be equal to the backfill internal friction angle ϕ_g^g.

In the DEM approach, the backfill transfers force to the walls by means of the block–grain contacts. Specific properties for such contacts must therefore be identified. A linear elastic contact associated with a Coulomb's law is adopted and the normal and tangential stiffness of the block–backfill grain contact was taken to be equal to 5.10^7 Pa, as indicated in section 1.2.2.2. Coulomb's law involves a local friction angle ϕ_i^l that must be identified. The calibration of ϕ_i^l was performed by modeling a typical soil–structure interface test and using a trial-and-error method.

More precisely, a constant normal load (CNL) test was modeled and the device together with the conditions at the boundaries for the test is given in Figure 1.14. The blocks were not entirely modeled and just a continuous surface modeling the outer shape of the wall blocks in contact with the backfill was considered (thin and long layer of particles at bottom in Figure 1.14). Any discontinuity existing between blocks was removed for the sake of simplicity. Nevertheless, it was checked that these discontinuities were out-of-scale entities that did not specifically contribute to the development of further shearing resistance in the interface.

The upper part of the system in Figure 1.14 consists of a box with rigid walls in which grains from the backfill have been poured under gravity. An initial porosity of 0.22 for the sample was warranted and if the local friction angle ϕ_g^{l} was reduced to reach this porosity, this angle was set back to 45° at the end of the process. A rigid lid closed the box and a constant vertical stress measured on this platen (equal to 35 kPa) was maintained throughout the CNL test. No friction was applied between the backfill particles and the rigid walls of the box. The lateral walls of the box being fixed, a horizontal velocity of 5.10^{-4} m/s was applied to the bottom system that modeled the DSRW surface. The size of the sample was 0.90 m in length and 0.50 m in height and composed of 925 backfill particles. The size of the box was roughly half the height of the DSRWs tested by Colas *et al.* [COL 10a] and was large enough to provide a sufficient number of contact points between the backfill grains and the simplified model of the wall.

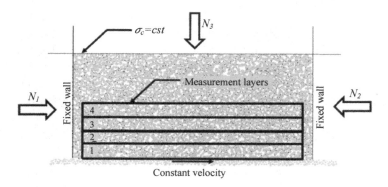

Figure 1.14. *Model for the CNL direct shear test for the backfill wall interface*

In an actual CNL test, the measurements are performed at a global scale due to a device being located outside the sample. Thus, there is no direct access to phenomena taking place at the backfill–wall interface and an inverse method is required to compute the stress field in it. More details about the methodology required can be found in [PRA 13]. Contrary to actual experiments, the DEM gives direct access to this information.

The lower part of the sample was then divided into four layers of $3D_{50}$ thick each with a full-length discarding the zone close to the vertical walls. In each layer, a stress tensor was computed from local information by a homogenization technique [ITA 08] and the mobilized friction angle in each layer was then deduced. Since the sample was actually one realization of a random process, four shear tests were simulated for a given value of the local friction angle ϕ_i^l. Then, for each relative displacement between the sample and the model for the wall, an averaged mobilized friction angle was computed.

The final result for a CNL test had three different values for the local friction angle ϕ_i^l: 0, 11 and 25° in Figure 1.15, where "In" denotes a result in layer 1 (Figure 1.14) and "Out" denotes the final result computed at the global scale (as performed in actual experiments). The 0° friction angle corresponds to a case where the geometrical roughness only contributes to the resistance of the interface. The corresponding curve indicates that shearing in the interface mainly came from the geometrical roughness and little from the value of the local friction angle ϕ_i^l. Indeed, a mobilized friction angle of 33° was found at peak in this case, which is a rather high value for an interface internal friction angle. In Figure 1.15, we can note that the "Out" curve, corresponding to a computation at the global scale tends to deviate from the "In" curve computed in layer 1 when ϕ_i^l is different from zero [ZHA 07]. Further blockages within the sample may contribute to this difference that may not exist when ϕ_i^l is equal to zero. Since local results are more valuable for characterizing the behavior of the interface, information taken in layer 1 is considered in the following for identifying ϕ_i^l. Finally, a local friction angle of 25° seems correct to retrieve a global friction angle of 37.7° for the backfill–wall interface.

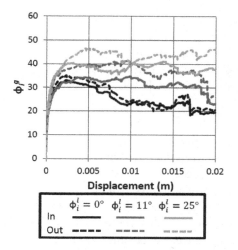

Figure 1.15. *Backfill–wall interface mobilized friction angle throughout the simulation of a CNL test*

Subsystems	Global	Local
Wall	$\phi_b{}^g$ = depending on the material, see Table 1.2	$\phi_b{}^l = \phi_b{}^g$
Backfill	$\phi_g{}^g$ = 37.7° $\phi_{gc}{}^g \approx 32°$	AR = 2.5 $\phi_g{}^l$ = 45° n_0 = 0.22
Interface	Rough interface → $\phi_i{}^g$ = 37.7°	$\phi_i{}^l$ = 25°

Table 1.1. *Local and global parameters for the three subsystems wall, backfill and backfill–wall interface*

Material	Contact friction angle
Schist	25°
Limestone	35°

Table 1.2. *Values of the block–block friction angle $\phi_b{}^g$ for the different stone wall materials*

The local parameters for the different subsystems, wall, backfill and backfill–wall interface, are summarized in Table 1.1. The values corresponding to the friction angle between blocks and corresponding to different materials used in the full-scale experiments by Villemus *et al.* [VIL 07] or Colas *et al.* [COL 10a] are given in Table 1.2.

1.2.2.4. *Modeling of slope DSRW's behavior*

The modeling of the full-scale experiments carried out by Colas *et al.* [COL 10a] first requires the construction of the wall. To achieve it, the wall blocks are regularly constructed with a smooth-joint contact law at their interface. The contact plane orientation follows the initial joint inclination α, i.e. collinear with α for the block–block interface between successive layers. This angle also corresponds to the batter of the outward wall face. Remember that the block model is rigid and the block is transformed into a hollow block, keeping the same mass. The size of the blocks is 33 cm in length and 12 cm in height and each hollow block consists of approximately 170 disks. The foundation of the wall is modeled by a single layer of fixed disks and the friction angle between the first layer of wall blocks and the foundation is taken to be equal to the friction angle of a block–block interface.

The backfill is then created layer by layer with a thickness of 15 cm until mid-wall height and then with a thickness of 10 cm, which is close to a block height. From a system at equilibrium where the backfill slope is equal to 32° (Figure 1.16(a)), which is the angle of repose for the backfill in full-scale experiments, a subsequent layer is created in the following way. Grains are poured from a container 2.5 m long with a drop height constant for all grains. Thus, the container is inclined from the horizontal with an angle of 32° (Figure 1.16(b)). The local parameters deduced in the previous section were used here (Table 1.1). The grains roll along the slope and when the layer in contact with the wall reaches the desired thickness, the local friction angle ϕ_g^l between grains was set to 15.6°. This allowed us to obtain a final porosity for the layer of 0.22. Equilibrium was then computed. Finally, the local friction angle was reverted to 45°.

The temporary reduction of the local friction angle modified the backfill slope to a smaller value to that desired. The slope was modified by removing particles before a further loading stage to achieve the expected value, i.e. 32°.

The disruption of the stability of the wall (while searching an equilibrium for it) due to the increase in loading was then checked according to two criteria. When the backfill height was less than 1.5 m (approximately half of the wall height), the whole system (wall+backfill) was classified as stable when the average unbalanced force was found to be smaller than 5.10^{-2} N. When the backfill height was greater than 1.5 m, a more restrictive average unbalanced force was used (1.10^{-2} N). A less restrictive stability criterion was considered during the first part of the simulation to reduce the simulation time since the failure of the wall is generally obtained after 17 days of computation (Intel Xeon CPU 3.2 GHz).

A second criterion denoted *failure criterion* is checked when the stability criterion is not satisfied. This latter criterion is based on the kinetic energy evolution of the very wall. If the wall's kinetic energy during 10 successive measurements (spaced by 500 time steps) is greater than 10^{-1} J, the wall is supposed to fail and the simulation is stopped. In fact, since the equilibrium for the wall is computed when a whole subsequent layer is created and stabilized, a dramatic rise in the kinetic energy of the wall is first observed. This energy tends to quickly decrease if the equilibrium can be found. A specific study was performed to check that 10 successive measurements were large enough if the equilibrium was supposed to be reached.

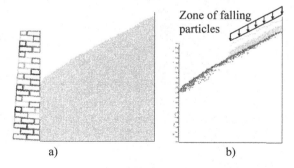

Figure 1.16. *Creation of a subsequent backfill layer on the top backfill: a) existing backfill and b) pouring of grains from a container. For a color version of the figure, see www.iste.co.uk/vincens/drystone.zip*

The kinematic velocity field for the wall blocks and the backfill grains for wall C2s tested by Colas *et al.* [COL 10a] is shown in Figure 1.18. In this figure, a toppling mode of failure for the DSRW is clear where an upper part of the wall overturned (with a merely monolithic movement), while the bottom part of the wall mainly remains fixed. The failure surfaces that cross the whole system are remarkable: they consist of two planes crossing the backfill and the bottom wall with different angles to the horizontal.

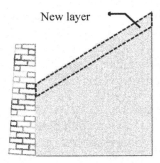

Figure 1.17. *Final state for a subsequent layer on the top backfill. For a color version of the figure, see www.iste.co.uk/vincens/drystone.zip*

Figure 1.18. *Toppling failure of wall C2s; kinematic velocity field. For a color version of the figure, see www.iste.co.uk/vincens/drystone.zip*

Table 1.3 shows the critical backfill heights found through DEM and the associated mode of failure for the three wall tests carried out during the full-scale experiments. For each wall, two simulations were performed, since the result is likely to be sensitive to the random process inherent to the creation of the backfill. We can note that the critical backfill height for cases C3s and C4s was greater than the wall height itself. A rigid column of disks was created at the top-inward wall face and fixed to this wall face. On site, a device was accordingly added to increase the wall height (small sheetpiles). Moreover, the results for computations are given with a precision of +/− 5 cm, which corresponds to half the thickness of a numerical backfill layer.

A relative error smaller than 7% was found in all cases between the results of the DEM and the on-site measurements. This result tends to validate both the methodology for obtaining the local parameters involved in this fully DEM study and the whole simulation process, including the creation of the wall and the backfill. The two computations that were performed for each wall tested allow us to estimate the expected precision of the results. Here, the results can only be given with a precision of 10 cm, which is greater than the precision of the computation due to the fact that the backfill is created by successive layers of 10 cm-thick backfill. Finally, the displacement field for three different stages of loading (backfill height h_s) is given in Figures 1.19(a) and (b). Figure 1.19(a) relates to the DEM simulation, while Figure 1.19(b) shows the measurements performed throughout the experiments. The results are fairly comparable for h_s smaller than 2.1 m. When the system is close to failure (h_s equal to 2.4 m), the discrepancy is important, but at this stage the system is very sensitive to the load increment where very large displacements can be observed.

Wall height h (m)	2.5	2.5	2.5
Wall label s: schist, l: lime	C2s	C3s	C4l
Experiments	2.3S/T	2.78T	2.72T
DEM	(1) 2.46T	(1) 2.68T	(1) 2.53T
Computation (1) and (2)	(2) 2.38T	(2) 2.72T	(2) 2.63T
Relative error	(1) 7%	(1) 4%	(1) 7%
	(2) 3%	(2) 2%	(2) 3%

S: sliding, T: toppling

Table 1.3. *Critical backfill height from DEM simulations and full-scale experiments*

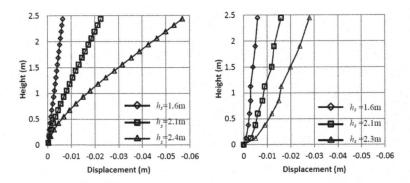

Figure 1.19. *Comparison of horizontal displacement profiles for wall C2s for different backfill height h_s: a) DEM modeling and b) experiment*

1.2.2.5. *Conclusion*

A full DEM approach to studying the failure of slope DSRWs was presented in this section. The walls were loaded by a backfill and the critical backfill height was measured. The technique to create all the individual objects involved in the whole modeled system and the process to identify the local parameters were given in detail. The results provided by the DEM computation were found to be in a very good agreement with the results obtained on site. The toppling mode of failure, which was found to be predominant on site, was also obtained through the DEM simulations.

The DEM was then proven to be able to model the 2D behavior of plane slope DSRWs accurately. It may be an oversized method, since time computations are very large; nevertheless, it can be considered as a reference for further numerical studies. Moreover, this method gives access to the whole deformation field throughout loading and before failure. This aspect is not necessarily important in the case of a slope-retaining wall where the maximum pressure induced by the slope is the most critical information for the design of such walls. This may not be the case for highway retaining walls, where deformations must be limited far beyond the critical loading induced by the vehicles. Indeed, too much deformation could induce damages to the road structure.

1.2.3. *A mixed DEM–continuum approach*

Due to the drawbacks of the full DEM approach – including high-computation time, heavy process to create the individual objects and identify the associated local parameters, as well as heterogeneities within the backfill and in the backfill–wall interface due to the use of elongated particles – we propose using a different approach to solve the problem of the stability of slope DSRWs. Nevertheless, it is desirable to keep the very nature of a DSRW: a discretized system able to dissipate a large amount of energy by friction and to ensure that large relative displacements between blocks can occur. Conversely, the heterogeneities within the backfill can be removed by using continuum modeling of both the backfill and the backfill–wall interface. This latter approach is expected to greatly reduce the number of degrees of freedom within the backfill, and as a consequence to greatly save computation time. In fact, modeling DSRWs by means of a discrete method for the wall and a continuum approach for the other subsystems was used by some authors to study some aspects of DSRWs [DIC 96, HAR 00, POW 02, CLA 05, WAL 07].

This section is devoted to the modeling of the full-scale experiments by Colas *et al.* involving plane-slope DSRWs using a mixed DEM–continuum approach. The results are compared to both what was observed on site and what was obtained by the full DEM approach that was previously presented.

Itasca code UDEC^{TM} (Universal Distinct Element Code) is used to model the loading of the slope DSRWs with a backfill. Other choices could have been made. In particular, a coupling between a DEM code and an finite element method (FEM) code could have been performed. Nevertheless, the exchange of information between two different codes is generally time-consuming since it involves the creation of numerous files. The UDEC^{TM} has the advantage of allowing the modeling of rigid or deformable interacting bodies with deformable contacts using an explicit solution method to solve the dynamic equations. The basic elements here are polygons that can be created giving the coordinates of their summits or by cutting a large body into several polygons by means of different cutting planes. The latter technique is used in the present work.

1.2.3.1. *Model geometry*

The wall is created by defining the global geometry of this subsystem and the use of cutting planes allows us to create both the blocks and the expected inclination of the layers (Figure 1.20). The blocks are considered deformable bodies according to Hooke's law and the contact between blocks is also deformable. As such, both a normal and a tangential stiffness must be defined. The contact is supposed to cause friction, and a homogenized Coulomb's law is used which requires the calibration of a frictional coefficient.

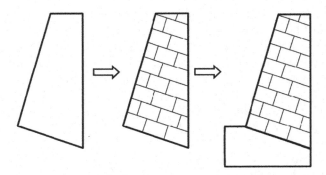

Figure 1.20. *Construction of the wall: general geometry, blocks and foundation*

The meshing of the whole backfill is built from the very beginning but not all the elements are activated for computation of the equilibrium of the whole system. The large bottom part of the backfill is first activated as a whole (Figure 1.21), since the equilibrium of the wall is not supposed to be broken for such a low backfill height. Then, 10 cm-thick layers are subsequently activated while the equilibrium of the whole system is computed to mimic the gradual loading of the wall on site. These layers have a slope corresponding to the slope found on site, i.e. 32°. A weightless block can be added at the top of the wall in order to model the role of an additional small sheetpile that was used on site to increase the wall's height by a small amount.

A non-associated elastic perfectly plastic Mohr–Coulomb model was chosen for the backfill. All the points at the backfill base are fixed and the contact between the backfill and vertical outer system can slide freely along the vertical direction.

The backfill–wall interface behaves as an elastic perfectly plastic interface for which internal friction angle and cohesion if required are parameters that must be identified. All the mechanical parameters of the three subsystems resulting from the calibration process are given in Table 1.4.

In fact, all the constitutive laws involved in this method being homogenized laws, the model parameters that are to be identified are directly derived from global measurements at the scale of the representative elementary volume.

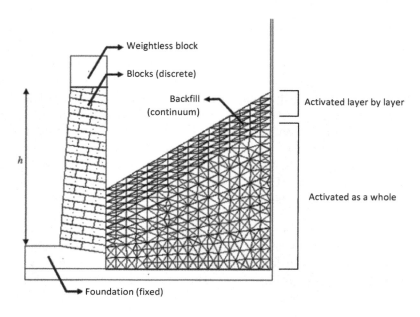

Figure 1.21. *Construction of the backfill*

Subsystems	Global
Wall	Normal and tangential stiffnesses at contact = 10^8 Pa/m ϕ^b_g = depending on the material, see Table 1.2
Backfill	Shear modulus: G= 38 MPa Bulk modulus: K = 55 MPa ϕ^g_g = 37.7° Dilatancy angle: Ψ=8°
Interface	Normal and tangential stiffnesses = 10^8 Pa/m Rough interface: ϕ^i_g = 37.7°, C= 0kPa

Table 1.4. *Global parameters for the three subsystems: wall, backfill and backfill–wall interface*

1.2.3.2. *Computation of the critical backfill height*

To determine the stability of the wall, a sole criterion based on the kinematic energy is used. Up to a backfill height of 40% of the wall height, the wall is supposed to be stable if the instantaneous kinematic energy of the wall is smaller than 1.10^{-2} J (a criterion that is not severe). For a backfill height that is greater than 40% of the backfill wall height, the criterion is restricted to a value smaller than 1.10^{-3} J in order to obtain a more precisely balanced system. If the criterion is reached, a subsequent backfill layer is activated. If the wall kinetic energy during 10 successive measurements (spaced by 1,000 time steps) is greater than 10^{-2} J, the wall is supposed to have failed and the numerical simulation is stopped.

A typical result for a wall failure (wall C4l in this case) is given in Figure 1.22. The critical backfill heights found through the numerical simulations for the different cases studied by Colas *et al.* are given in Table 1.5. We can note that the quality of the prediction of the critical backfill heights is very close to the one found by the full DEM approach. The gain provided by the mixed DEM–continuum approach is huge since a computation for a wall case is about 4 h, while in section 1.2.2.4, there was a computation time of about 2.5 weeks.

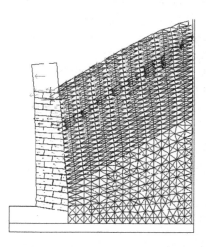

Figure 1.22. *Simulation of the failure of a slope DSRW C4l. For a color version of the figure, see www.iste.co.uk/vincens/drystone.zip*

Wall height h (m)	2.5	2.5	2.5
Wall label s: schist, l: lime	C2s	C3s	C4l
Critical backfill height: – Experiments (m)	2.3S/T	2.78T	2.72T
– DEM–continuum approach (m)	2.41T	2.62T	2.52T
– Relative error	5%	6%	7%

S = Sliding T = Toppling. Critical backfill heights are given with a precision of ±5cm.

Table 1.5. *Critical backfill height from DEM–continuum numerical simulations and full-scale experiments*

1.2.3.3. *Parameters influencing the stability of a slope DSRW*

A parametric study was carried out to better analyze the main properties of the overall system that essentially lead to its failure and resistance. More precisely, the question that arises is whether an error in the calibration of the model's properties has the same impact on the wall's stability irrespective of the model parameter considered. A first study considered the influence of an error equal to $\pm 3°$ on the average block–block contact friction angle. The simulations showed that the departure for the backfill critical height is almost insignificant ($< 1\%$). It means that the departure is actually within the range of the error due to the thickness of the incremental backfill layer considered, i.e. ± 5 cm. An error equal to $\pm 3°$ for the backfill–wall interface friction angle induced a departure of $\pm 5\%$ with respect to the reference cases investigated in section 1.2.2.4. Finally, the same departure was obtained if an error of $\pm 3°$ was considered for the backfill internal friction angle.

At this stage, we can consider that correctly identifying the properties of the backfill seems to be more important than a precise characterization of the frictional property of contacts between the wall blocks. The pattern related to the toppling mode of failure where an overturning moment is at stake may explain this.

The last parameter that greatly influences the stability of the wall and its design is a property of the backfill that was not considered in the full-scale experiments by Colas et al. [COL 10a], i.e. cohesion. As an illustration, a case study is presented hereafter. For the sake of simplicity, the slope of the backfill surface is taken to be horizontal, a wall and a backfill height equal to 2.5 m are chosen with an apparent specific weight for the wall equal to 2,050 kg/m^3, a friction angle at block contact of 25°, an apparent specific weight for the backfill equal to 1,520 kg/m^3, an internal friction angle for the backfill of 30° and a backfill–wall friction angle of 30°. First, a reference case is computed in order to design a wall that can just handle the loading coming from the backfill with a cohesion C equal to 0 kPa (purely frictional material). Then, two cases are investigated where the backfill cohesion is equal to 2 or 5 kPa. In each case, the geometry (wall base dimension) is computed again so that the weight of the wall can just balance the driving moment due to the backfill loading.

The result of this study is given for a wall unit length in Table 1.3. We can note that considering a cohesion as small as 2 kPa leads to an increase of 37% in the volume of the wall material, which is substantial. A cohesion of 5 kPa can reduce the volume of the wall material by 74% compared to the case where C is equal to 0 kPa. In a more illustrative way, a safety factor against overtopping failure is computed if the cohesion is not taken into account in the design of the DRSW. The safety factors that were obtained are also given in Table 1.6. The overdesign of the wall is very important and becomes even more excessive as the cohesion increases.

Backfill cohesion C (kPa)	0	2	5
Base wall dimension (m)	0.528	0.363	0.191
Wall weight (t/m)	2.35	1.49	0.6
Relative weight difference with case C=0kPa	0%	37%	74%
Safety factor against failure	1	2.2	9.4

Table 1.6. *Base wall dimension and gain for a unit wall length if the backfill cohesion is taken into account*

1.2.3.4. *Conclusion*

A mixed DEM (for the wall) and continuum approach (for the backfill) is used for the study of slope DSRWs. This approach holds many advantages; the first advantage is that the quality of the prediction is very close to the more sophisticated full DEM approach. The second advantage is that the computation effort is far lower than in the case of the full DEM approach. A result for the critical backfill height is provided here within 4 h, which is acceptable from an engineering perspective. Since all the constitutive laws considered are homogenized laws (including block–block contact law), the parameters involved in the models are associated with a scale greater than the representative elementary volume and can be calibrated easily.

Concerning the block geometries that can be handled, only polygonal bodies can be created with UDEC, which is not a restriction in the case of DSRWs. In the case of slope DSRWs, a study showed that the block shape and dimensions play a minor role in the justification of the stability of a slope DSRW where mainly friction between blocks and the overall weight of the wall drive the resistance of the wall against failure.

1.2.4. *Conclusion*

Failure of plane slope DSRWs occurs according to a sliding mode or a toppling mode with a plane deformation state. A top part of the wall moves with respect to a bottom part of the wall, with an almost rigid displacement mode. The toppling mode of failure was the predominant mode observed through full-scale experiments.

In the case of plane slope DSRWs – the mode of failure involving a plane strain condition – a 2D analysis may be suitable to address the behavior of such a system. For DSRWs that would hold a curved shape, a 3D analysis is required.

Different techniques have been used to model the behavior of plane slope DSRWs when they are about to fail. Three subsystems are involved in the modeling: the wall, the backfill and the wall–backfill interface. First, a full DEM approach was used to model the whole system. In this case, the creation of individual bodies requires particular care and the calibration of the local parameters is processed using a trial-and-error method. The method is proven to give a very good prediction of the critical backfill height that triggers failure. It can be considered as the most sophisticated method for modeling the actual behavior of slope DSRWs. It has the advantage of allowing the modeler to control the phenomena at stake in the three subsystems in more detail. Moreover, it gives access to both the stress and the deformation fields within the overall system. Nevertheless, the calibration of the parameters requires considerable effort and the overall computation is very time-consuming.

An alternative to this method is the use of a mixed DEM–continuum approach, where some advantages from the full DEM approach are conserved and where the drawbacks are mitigated. The idea is to keep the discrete nature of the wall where large relative displacements between blocks can occur and to decrease the number of degrees of freedom of the whole system by modeling the backfill as a continuum medium. This method gives excellent results with a very good estimate of the critical backfill height triggering failure, while the computation time is dramatically reduced. The duration of the study involving the creation of the geometry of the whole system, the identification of the model parameters and the computation make this approach appealing for the engineer.

The mixed DEM–continuum approach allows us to have access to the deformation field within the wall. This possibility is not allowed with some simpler methods such as the yield method [COL 10b]. Moreover, detailed parametric studies can be carried out. Some parametric studies showed that in the case of a failure by toppling, which is the general case for plane slope DSRWs, the uncertainties related to the determination of the interface friction angle and the backfill internal friction angle have a similar impact on the determination of the backfill critical state height. The uncertainty of the block–block friction angle seems to have a lesser impact for this kind of failure.

Moreover, an uncertainty in the determination of the porosity of the wall (±5%) has the same impact on the determination of the backfill critical state height as an uncertainty of ±3° for the backfill–wall interface friction angle and the backfill internal friction angle for the friction angles investigated. This range of uncertainties being representative of uncertainties on site, special care is required for this kind of construction in order to provide a reproducible compactness for the wall.

So far, the cohesion of the backfill is not taken into account in the current French recommendations for the design of plane slope DSRWs. Nevertheless, some ongoing updates will include this perspective that allows a definite gain of material for the wall construction. For example, we found that for a horizontal backfill slope and a wall of 2.5 m, taking into account a cohesion of 2 kPa, allows 37% reduction in block volume, which is considerable. If the cohesion is equal to 5 kPa, the gain increases to 74%.

1.3. Highway dry stone retaining walls

1.3.1. *Experimental campaigns*

Highway DSRWs are structures for which an extra concentrated loading applied at the backfill surface complements the direct loading on the wall coming from the backfill itself. This concentrated loading simply models the weight of the vehicles (transferred to the road by the contact between the wheels and the road surface) operating on the road.

These experiments have been carried out using small clay bricks with interlocking comparable to that on site with actual rubble stones. The wall bricks were 88 cm wide, 3.3 cm thick and 16 cm high. The dimensions of the bricks in the model were 33, 16 and 11 mm, respectively. For the small-scale model, an arrangement of blocks formed by a header and two blocks in stretcher was chosen. This arrangement is the basic cell that can be found in actual DSRWs. A Hostun sand was used for the backfill. The box containing the overall system was 110 cm in width, 50 cm in depth and 40 cm in height.

The concentrated loading was modeled by steel blocks (6 cm by 7 cm wide) placed on the backfill ground surface. The horizontality of the first block that will bear subsequent steel blocks was carefully controlled. Different distances d from the inward wall face were studied: 2, 3 and 4 cm. Successive blocks were superposed on the primary steel block until the wall collapsed. For each distance d of the steel block from the inward wall face, five tests toward failure were performed. Further details about this small-scale experiment can be found in [QUE 15]. The results are given with a precision of 1N due to the use of steel units of 0.1 kg. A sketch of the system after failure is shown in Figure 1.23.

Figure 1.23. *View of the experimental setup-up after failure, d = 4 cm*

1.3.2. *Mixed DEM–continuum approach*

Since failure in this case appears with true 3D deformations, 3DEC (Itasca code) software was used to model the small-scale experiment. This code allows the equilibrium of deformable polyhedral bodies in interaction to be computed.

1.3.2.1. *Creation of the system and loading*

The approach considered herein is similar to what was presented in section 1.2.3. The wall is supposed to be made up of individual bodies, while the backfill and the wall–backfill interface are modeled as continuum media.

The wall was created from a single block that was eventually split by different cutting planes in order to generate the parallelepiped bricks. The backfill was created in a single stage and the meshing size of the backfill was reduced within the zone where the concentrated loading was placed (Figure 1.24). Large shear strains were likely to occur in this zone when the system was getting close to failure due to the intensity of the concentrated loading F. The reduction in the meshing size allowed eight gridpoints of the backfill to be involved in the transfer of forces to each wall brick within the wall–backfill interface, which was satisfying. More details can be found in [QUE 15].

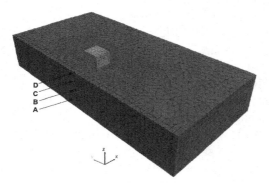

Figure 1.24. *DSRW modeling with a concentrated load on the backfill surface; A, B, C and D are observation points. For a color version of the figure, see www.iste.co.uk/vincens/drystone.zip*

The first layer of bricks (bottom wall) was fixed as in the experiment and the lateral sides of the retaining walls were only allowed to move vertically. For the backfill, in the experiments, the material at the boundaries was in contact with the wall of the box. In the numerical simulations, the associated gridpoints on the external vertical faces were only allowed to move vertically, while at the basis of the backfill the gridpoints could only move horizontally.

The wall bricks were supposed to follow Hooke's law and the joint between the wall bricks was ruled by a Coulomb slip model where the incremental law for both the normal force and the tangential force was linear. Coulomb's law allowed the tangential force to be limited to a given value related to the frictional property of the joint. The behavior of the backfill was modeled by the elastic perfectly plastic Mohr–Coulomb constitutive law and the joint between the wall and the backfill was addressed in the same way as a joint between blocks.

The elastic properties of the blocks were is taken from the literature but the brick–brick friction coefficient was identified by a test on an inclined plane involving several bricks. The properties of Hostun sand were taken from [FLA 90]. For the sake of simplicity, the usual dependency of the elastic moduli with depth is not taken into account. In this experiment, the mean pressure in the backfill at wall mid-height is equal to 0.7 kPa, which is far smaller than the one considered in the works that defined the dependency of Young's modulus with confining stress [BIA 94, DAN 02]. The relationship proposed by Biarez and Hicher [BIA 94] was nevertheless used to identify the Young's modulus for an average pressure of 0.7 kPa. Moreover, since the sand was poured at almost zero drop height, a very loose state was found for the material (relative density of 2%) and thus no dilatancy is considered in the simulations. In this case, the internal friction angle was equal to the critical friction angle. A test aimed at identifying the angle of repose for the material which was expected to be close to the critical state angle confirmed the critical state angle of 32° found by Flavigny et al. [FLA 90] for a confining stress of 50 kPa.

For the wall–backfill interface, the friction angle was taken as being equal to two-thirds of the backfill internal friction angle. This relationship is the usual one taken for precast concrete face of the walls in soils. The associated relative roughness seemed to be representative of the one existing for the clay wall brick–sand interface. The mechanical parameters identified for the different subsystems are given in Table 1.7.

Some preliminary calculations showed that the perfect contacts between the wall bricks tended to generate jammed states throughout the loading process and no clear steady state could have been observed for the concentrated loading. Details about this feature can be found in [QUE 15]. In fact, these perfect contact planes did not exist in the down-scaled experiments due to imperfections in the clay bricks. Simulations proved that a gap of 0.5 mm (5% of brick thickness) between the vertical contact planes was enough to mitigate this bias. Above this threshold value, the result became independent of the gap value.

Subsystems	Wall bricks	Backfill	Interface
Specific weight (kg/m³)	1,635	1,300	-
Bulk modulus (Pa)	5.56×10^8	8.33×10^6	-
Shear modulus (Pa)	4.17×10^8	3.85×10^6	-
Internal friction angle (°)	32	32	20
Cohesion (kPa)	0	0	0
Dilatancy angle (°)	0	0	0
Normal stiffness (Pa/m)	1×10^9	–	1×10^9
Tangential stiffness (Pa/m)	1×10^9	–	1×10^9

Table 1.7. *Mechanical parameters for the three subsystems: wall, backfill and backfill–wall interface*

1.3.2.2. *Characteristics at failure*

The results for the maximum vertical concentrated force on the backfill surface throughout experiments and throughout the simulations are given in Figure 1.25. We can note that the dispersion of experimental results increases with d value. It is mainly due to the uncertainties related to the homogeneity of the backfill. Indeed, as the distance between the steel block and the inward wall face increases, the volume of sand involved in the transfer of stresses toward the wall increases as well. Thus, more numerous heterogeneities within the backfill may impact the value of the concentrated force at failure more easily.

The maximum force F_{fail} triggering failure throughout the simulations is also given in Figure 1.26 and is generally in very good agreement with the mean value for F_{fail} (denoted by $\bar{F}_{fail}$) found throughout the experimental tests. Nevertheless, a slight departure from $\bar{F}_{fail}$ for d equal to 2 cm was observed. It may be due to the limit of the continuum approach related to a detachment of the backfill from the wall in the simulation at the head of the wall. This detachment is a consequence of punching of backfill by concentrated loading. Even if observed for any value of d, this detachment appeared in an early loading stage for d equal to 2 cm. For d equal to 3 or 4 cm, this detachment was only observed in the later loading stages, i.e. very close to failure, and has little impact on the final value F_{fail}.

The maximum departure of the simulated results from $\bar{F}_{fail}$ is equal to 8%, found in the case when the steel blocks are very close to the wall.

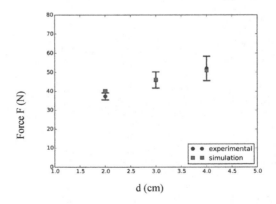

Figure 1.25. *Maximum vertical force on the backfill surface inducing total wall failure for different distances d of the steel blocks from the inward wall face*

Figure 1.26 gives some clues about the reason for the increase in F_{fail} with d. When the concentrated loading is close to the wall (the wall is hidden at the left side of the system), the loading on the wall affects the upper part of the wall, which contributes to the driving forces that are able to destabilize the wall more easily. When the loading is further away from the wall, zones of the wall located closer to the bottom are impacted and the contribution to the driving forces destabilizing the system is reduced. It implies that a greater value of concentrated force is required to induce failure. This conclusion can also be drawn by considering the bulbs of iso-stress values in the framework of Boussinesq theory.

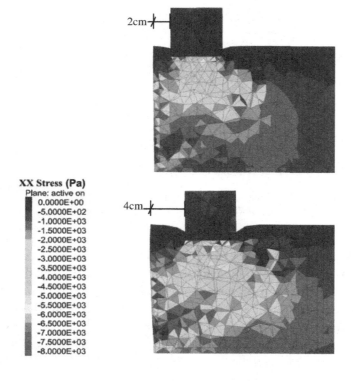

Figure 1.26. *Horizontal stress field (σxx) as a function of the distance between the wall and loading block, for the final stage of loading: (a) d = 2 cm and (b) d = 4 cm. For a color version of the figure, see www.iste.co.uk/vincens/drystone.zip*

Figure 1.27 plots the evolution of the horizontal displacement of point D (at the head of the wall) when the concentrated force increases. All the curves exhibit the same general features: first, point D hardly moves until the concentrated loading F reaches a certain value denoted herein by F_{crit}. For larger values of F, large displacements take place and F evolves toward a plateau which corresponds to the maximum force the wall can bear.

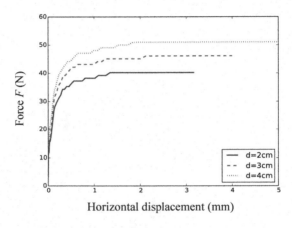

Figure 1.27. *Force–displacement curves as a function of the distance d between the loading steel block and the inward face of the wall. Displacements are measured the head of the wall (point D in Figure 1.24). For a color version of the figure, see www.iste.co.uk/vincens/drystone.zip*

A possible definition for F_{crit}, which can only be a conventional one, was found by monitoring the value of the incremental shear rate, defined as the ratio between the brick velocity (point D) and the wall height. In Figure 1.28, we can note that the shear rate shows a steep inflexion, for any value of d, when the value 0.1 m.s^{-1}/m is exceeded. It means that for greater values of concentrated force, the system overreacts and large displacements are likely to take place. A related displacement u_{crit} at point D can be computed on the basis of Figure 1.27. The values for u_{crit} are given in Table 1.8, together with the results concerning the forces at failure for both the experiments and the simulation and the force at the critical stage.

We can note that F_{crit} tends to get closer to F_{Fail} when d becomes small. It means that, in this case, the behavior of the system resembles that of a rigid perfectly plastic system.

Finally, Table 1.8 gives the value of the maximum length of influence l_i of the concentrated loading on the wall, computed at the head of the wall and lengthwise. The displacement of the bricks due to concentrated loading is supposed to be insignificant when the ratio between the total horizontal displacement of a brick and the wall thickness (i.e. 33 mm) is smaller than 1%. This distance is related to the zone of the wall which is likely to be affected by damage, if any, due to the existence of concentrated loading on the backfill.

As an illustration, Figures 1.29(a) and (b) show the magnitude of displacements within the backfill and the wall for the case of d equal to 4 cm during the last stage before total failure. In particular, Figure 1.29(a) gives a general view where the maximum influence distance of the concentrated load on the wall (measured at the head wall and lengthwise) l_i is reported. This distance, which is equal to 67 cm in this specific simulation, is found to be in fairly good agreement with the observation throughout the small-scale experiments, since in this case a value of 68 cm was reported (Figure 1.23). Distance l_i generally depends on the value of d (Table 1.8).

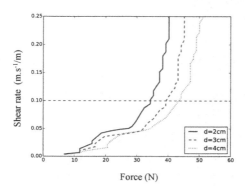

Figure 1.28. *Shear rate at point D for a possible definition of F_{crit}*

Distance d (cm)	2	3	4
$\overline{F}_{fail}$ (N) experiments	37.4	46.0	52.0
F_{fail} (N) simulations	39.2	46.1	56.9
u_{crit} (mm)	0.31	0.33	0.43
F_{crit} (N)	34.3	39.2	44.1
l_i (cm)	58	63	67

Table 1.8. *Characteristics of the system throughout loading for different distances between the concentrated loading and the inward wall face*

Figure 1.29(b) clearly reveals the zone within the backfill affected by failure. We can note that some backfill elements are not in contact with the wall's inward face. This feature appears when the system is very close to failure and consequently does not affect the value of maximum concentrated loading.

1.3.3. *Conclusion*

A mixed DEM–continuum approach was used in order to model a small-scale experiment where the wall was built with parallelepiped clay bricks. The numerical model was able to retrieve the mean maximum concentrated loading on the backfill surface triggering failure. This validation involves different distances loading from the wall's inward face.

The simulations show that as concentrated loading gets closer to the wall's inward face, the impact of the loading on the wall is concentrated on a reduced surface and on the upper part of the wall. The contribution of these stresses to the driving forces, inducing failure, is then important in this case.

The limit of the numerical work lies in the constant value for Young's modulus in the backfill. In the authors's opinion, the usual dependency on the square root of the effective average pressure may not be adapted for the range of stresses involved in the experiments. Further computations have been performed for a Young's modulus of 5 and 15 MPa and the results can be found in [QUE 15]. Except for the case of d equal to 2 cm, the results for the force at failure were found to be in the range of uncertainties throughout the experiments.

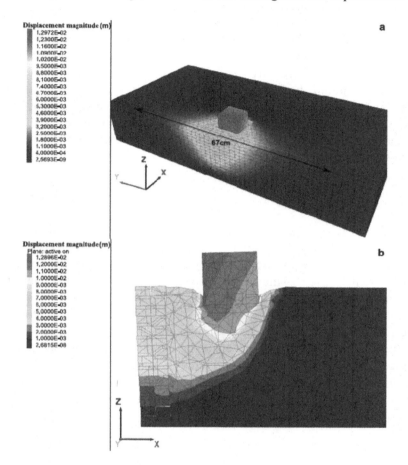

Figure 1.29. *Displacement field for the backfill and wall, d=4 cm: a) general view and maximum influence length of the concentrated loading: and b) cross-view along the plane of symmetry of the whole system. For a color version of the figure, see www.iste.co.uk/vincens/drystone.zip*

It seems too early to extrapolate these results to actual highway DSRWs since due to the thickness of actual walls, the arrangement of blocks can be more complex than considered herein. Moreover, the experiments were carried out using very standardized parallelepiped bricks that do not reflect the shape of actual rubble stones used on site for DSRWs. Nevertheless, the main features observed in these scaled-down experiments are expected in actual DSRWs.

Finally, the design of a highway DSRW is supposed to involve a criterion related to a maximum deformation allowed for road structures and not necessarily a criterion in terms of forces. It means that a maximum horizontal displacement for the wall blocks is expected to be specified. In this case, the role of the arrangement and the shape of blocks are critical, and full-scale experiments are required to further validate the use of a mixed discrete–continuum approach to study the behavior of highway DSRWs.

1.4. Conclusion

The behavior of slope DSRWs when they approach failure has been modeled using a full DEM and using a mixed discrete–continuum approach; discrete for the wall and a continuum method for the backfill. These two methods were able to found the critical backfill height obtained throughout full-scale experiments with a departure smaller than 7%, which is considered as a good result. Even though the full discrete approach is very time-consuming and not accessible for engineers, it allowed us to chart the course for the modeling of a typical complex system that requires a rigorous method for the determination of local parameters. It can be considered as the most sophisticated approach to the problem and can serve as a reference for more straightforward approaches.

The mixed discrete–continuum approach is a first step toward simplification where the continuum approach allows us to decrease the number of degrees of freedom involved for the backfill. The quality of results obtained is of the same order as those provided by the full discrete approach, with a definite reduction in computation time.

At the same time, these two numerical methods allowed us to validate the full-scale experiments performed on DSRWs and an analytical method as well (the yield design method). This latter method is rustic and less precise than the numerical methods, however it facilitates the construction of charts for a simple design of slope DSRWs. Nevertheless, it cannot give access to the deformation field of the wall, contrary to the numerical methods.

The modeling of the behavior of highway plane DSRWs approaching failure was carried out using a 3D mixed discrete–continuum approach on the basis of scaled-down experiments. In this case, the loading is concentrated at the surface of the backfill. The numerical results were found to be promising for the use of this technique, nevertheless, the excessive punching of the backfill by concentrated loading leads to deterioration in the quality of the modeling of phenomena in the vicinity of the loading. This does not impact the resistance of the wall obtained at failure except when the loading is very close to the wall. In this case, a detachment of the backfill from the wall and at the head of the wall was observed in an early stage of loading, leading to a slight overestimation of the wall's resistance.

1.5. Notations

R_n wall-backfill normalized roughness;

AR grain aspect ratio;

ϕ_b^g block–block global friction angle;

ϕ_b^l block–block local friction angle;

ϕ_g^g backfill internal friction angle;

ϕ_{gc}^g backfill critical friction angle;

ϕ_g^l backfill local friction angle;

n_0 backfill porosity;

ϕ_i^g wall–backfill interface global friction angle;

ϕ_i^l wall–backfill interface local friction angle;

d distance between the edge of the concentrated loading and the inward face of the wall;

l_i distance of influence of the concentrated loading;

F_{fail} maximum concentrated loading corresponding to wall failure in the simulations;

$\overline{F}_{fail}$ average maximum concentrated loading corresponding to wall failure in the experiments;

F_{crit} critical concentrated loading above which large displacements of blocks are triggered;

u_{crit} critical displacement at the head of the wall corresponding to F_{crit};

1.6. Acknowledgments

The work presented in this chapter was part of research projects C2D2 10 and MGC S01 denoted by PEDRA (Behavior of dry-stone or weakly bonded structures) funded by the Ministry of Ecology (MEDDE) and the Civil and Urban Engineering Network (RGCU), RESTOR (Restoration of Dry Stone Retaining Walls) funded by the Ministry of Culture (MCC) under the PNRCC program and MapCoD (Materials and processes for sustainable constructions) funded by the French Agency of Research (ANR) under the Projet d'Avenir Lyon-St Etienne. The author wants to thank these institutions for their support.

The main results presented herein were obtained by J.J. Oetomo during his PhD thesis and by J-C. Quezada during a post-doctorate. The author is very thankful for the quality of the collaboration that provided the results presented herein. Finally, the author also wants to acknowledge the French Institute of Indonesia (Ministry of Foreign Affairs) for granting some extra funds for J.J. Oetomo's PhD thesis.

Rockfill Dams with Dry Masonry

2.1. Introduction

The work presented in this chapter is the result of a joint effort between a French research laboratory, the Laboratory of Tribology and Dynamic of Systems (LTDS) from Ecole Centrale de Lyon, and an operator of dams, EDF (Electricité de France) which has been ongoing for more than a dozen of years.

Under the supervision of Professor Cambou, the LTDS, after developing a series of powerful constitutive models for soils (CJS1 to CJS4), moved to particle modeling with the idea that granular soils would be more efficiently simulated by particle movements in accordance with simple interaction laws rather than complicated constitutive models. Professor Cambou got this idea after working with Professor Jean Biarez at the University of Grenoble and Raul Marsal in Mexico. In the 1950s, both of them launched the use of particle modeling to study the behavior of rockfill dams. Unfortunately, at that time there were no tools able to integrate such an approach. In the 1990s, the situation changed with the development of a commercial code PFC^{2D} (Itasca SAS), based on the discrete element method (DEM). Indeed, the first applications [MAH 96] showed that particle methods are a natural way to address problems with soil behaviors that involve large particles in narrow structures,

Chapter written by Jean-Jacques FRY and Jean-Patrick PLASSIARD.

large deformations and discontinuities, which are difficult to model using the conventional finite element method.

EDF is the owner of more than a dozen of rockfill dams, built between 1930 and 1960, with an average height of about 20 m. This type of dam has steep slopes, covered by a dry stone revetment. This design is no longer used. Safety assessments of such dams by LTDS and EPF address three issues. The first difficulty of the safety assessment is the impossibility of testing a full sample of rock, because the representative volume requires several dozen cubic meters of material. It explains why there is no consensus on how to evaluate the friction angle of this material. The second difficulty is that the dry stone pitching, consisting of up to four rows of stones, is especially challenging for traditional numerical methods. The final challenge of the joint research between LTDS an EDF was to obtain a clearer understanding of the displacements of some benchmarks positioned at the crest or on the slopes of actual rockfill dams. Indeed, they exhibit a constant deformation rate over time, though this rate is usually supposed to decrease in time. Meanwhile, a better assessment of the safety margin was expected to be provided.

Section 2.2 is devoted to a comprehensive assessment of the properties and behavior of these dams. EDF launched the ACABECE project, piloted by François Laigle and based on a continuum approach using FLAC software (Itasca code). However, EDF was motivated to test the DEM so it had another way of understanding the behavior of these dams and confirming the safety of this stock of strategic works.

The feasibility study was commissioned by EDF in 1996 and performed by Mahboubi, after his thesis [MAH 97]. It was demonstrated that DEM mechanics was able to simulate the particulate nature of rockfill and capture evolving failure mechanisms in the triaxial tests by simulating realist stress-strain and volume change curves. The real opportunity was given to LTDS through the MICROBE research project (2001–2004). This project developed a numerical method based on the DEM for simulating blocks of rock and dams. A first PhD student, Rémi Deluzarche, modeled the static and dynamic stability of a rockfill dam (2001–2004) using PFC2D. A second PhD student, Thi-HuongTran (2003–2006), with the participation of a masters student, Romain Venier (2005), modeled the long-term behavior and creeping of dams and made the first comparison between two-dimensional (2D) analysis and three-dimensional (3D) analysis in plane strain condition with PFC3D. Jean-Patrick Plassiard was in charge of completing the final procedures and simulations (2008–2009). Section 2.3 summarizes the main results of these simulations.

These advances in particle methods for rockfill modeling hold promise for a deeper insight into dam engineering applications. However, these simulations were not calibrated on enough measured properties. The validation was therefore incomplete due to the lack of physical experiments. The last project, called PEDRA (2011–2014), aimed to provide physical models for the validation of future numerical simulations. Section 2.4 reports the data resulting from the failure of a down-scaled (1/10) physical model of a rockfill dam.

2.2. Dam performance and rockfill behavior

2.2.1. *Development of the dry masonry face rockfill dam*

The development of the dry masonry face rockfill dam began around 2600 BC, at the beginning of the Pyramid Age [SCH 94] in Egypt. Sadd-El-Kafara dam was discovered near the capital city of Memphis. It is the oldest known water-retaining structure of such a size in the whole world, although Herodotus mentioned that the founder of the first dynasty of kings built one at Kokeish in about 2900 BC. The overdesigned cross-section of this latter dam has a height of 14 m and a crest width of 56 m (Figure 2.1), probably to retain the core of silty sand and gravel between the two shoulders of rockfill [GAR 85].

This first construction underlines the problem of watertightness and protection: to retain the rockfill material, 17,000 revetment blocks were placed on both outer faces of the dam, each weighting 300 kg. It is interesting to observe the state of the blocks of this dry masonry 4,600 years after completion (Figure 2.2): most of them are eroded, a few have completely disappeared; however, no stability problem was noticed in the past.

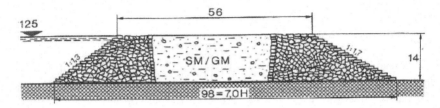

Figure 2.1. *Cross-section of Sadd-El-Kafara dam [GAR 85]*

Figure 2.2. *Upstream dry masonry of Sadd-El Kaffara*

This type of dam experienced a first comeback when the Californian mining industry developed its use during the gold rush of 1848–1849 in the Sierra Nevada Mountains. Experience up to 1960 using dumped rockfill demonstrated this type of dam to be a safe and economical, but subject to concrete face damage and leakage caused by the high compressibility of the segregated and loose dumped rockfill. As a result, concrete facing became unpopular, although rockfill had been demonstrated to be a high-strength and economical dam-building material. Partly in response to these problems, the earth or steel central core rockfill dam, with compressible dumped rockfill, was developed. The highest in the world at the time of its construction (1886) was the Otay Creek Dam [SCH 94]. It consisted of heaps of loose rocks sloping on both faces at their natural angle of friction with a central steel core. Other upstream-facing rockfill dams in the southwestern USA were made watertight by upstream decks of timber, steel or concrete. This allowed further reductions of their volume, since its mass now fully served to withstand the water load. The minimum volume was obtained by using reinforced rockfill, also called dry masonry, where blocks are placed by hand. The dry masonry permitted a higher bulk density, greater strength and more rigid structure. Different solutions were designed, mixing dumped rockfill zones with hand-placed rockfill zones [COY 39]. First, in the USA before 1900, the hand-placed rockfill zone was only below the upstream face, in order to reinforce the facing and to manage the scarcity of workers. In Italy, however, dry stone fill (20m<h<50m) was built with hand-placed rockfill in the 1920s (Figure 2.3) due to the low cost and high quality of workers.

French engineers developed the Italian design in Algeria in the 1930s at Foum-el-Gueiss (20 m), Bakhadda (45 m), Bou-Hanifia (54 m) and Ghrib (71 m) and in France on rocky foundation. In France, between 1940 and 1960, during and after World War II, a period during which human work was at low cost, 13 rockfill dams were built (Table 2.1), mostly based on the US design including the Italian steep downstream slope, protected by a dry stone pitching. In the 1930s, it was believed that the slopes of the dam could be independent of its height [COY 39] and the water load.

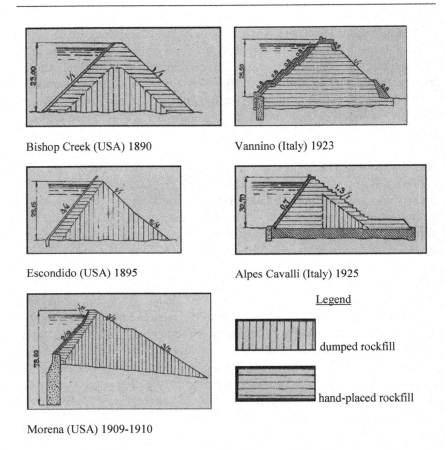

Bishop Creek (USA) 1890

Vannino (Italy) 1923

Escondido (USA) 1895

Alpes Cavalli (Italy) 1925

Legend

dumped rockfill

hand-placed rockfill

Morena (USA) 1909-1910

Figure 2.3. *Cross-sections designed in USA and Italy [COY 39]*

The main features of the rockfill dams built in France are small to medium height of less than 30 m (Table 2.1 and Figure 2.4) and challenging slopes, which are considerably steeper than the H/V = 1.4/1 slope of the current design (Figure 2.5). The crest width is small at around 2.5 m. Concrete face covered the upstream cemented masonry.

Since the advent of vibratory-roller-compacted rockfill in the 1950s, there has been continuous progress in aspects of design and construction methods. The slopes started to be gentler than they used to be, the stability was improved and the height was increased. Figure 2.4 illustrates the trends in the height of the concrete face rockfill dams (CFRDs) up to the year 2000. Today, the CFRD is a major dam type, spread over the world with a standard design.

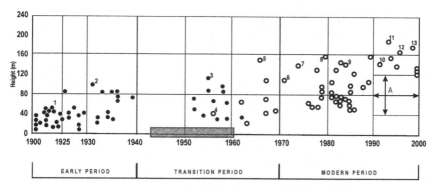

Figure 2.4. *Heights of CFRD dams versus time*
[ICO 07]; the rectangle dams built in France

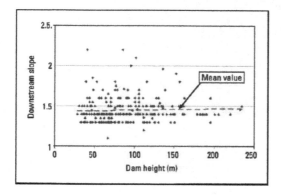

Figure 2.5. *Downstream slope versus height*
[FRO 09]; rectangle dams built in France

2.2.2. Cross-section, construction process and stone pitching

Figure 2.6 shows the cross-section of a typical dam, built with dumped rockfill on compressible foundation. Figure 2.7 shows the design of a typical dam, built with dry stone fill.

Date	Dam	Height	Date	Dam	Height
1942	Araing	25 m	1951	Chammet	19.3 m
1943	Laurenti	15 m	1951	Greziolles	30 m
1949	Grandes Pâtures	20 m	1952	Saint-Nicolas	6 m
1950	Auchaize	6 m	1953	Escoubous	20 m
1950	Portillon	22.5 m	1953	Les Laquets	13 m
1950	Vieilles Forges	10 m	1959	La Sassière	30 m

Table 2.1. *Rockfill dams built in France from 1940 to 1960*

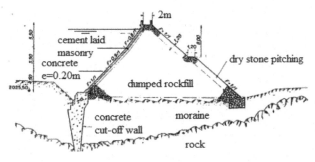

Figure 2.6. *Example of a dumped rockfill dam: photography of the downstream face and cross-section (Courtesy of EDF)*

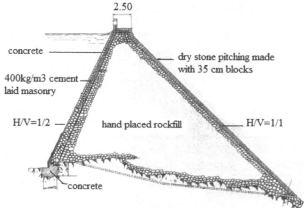

Figure 2.7. *Example of a typical rockfill dam built with hand-placed rockfill: photography of the construction and cross-section (courtesy of EDF)*

The international construction design at that time for dumped rockfill had three main features: selection of a quarry of hard and unaltered rock, dumping blocks of various dimensions in the dam body, and sluicing large quantities of water in the rockfill. Without this precaution, high settlements occurred following strong rainfalls (1 m at the 100 m-high San Gabriel Dam N°2 following the first storm and 3 m following the second strong storm). However, there is no evidence that the construction process followed all these requirements. The main requirement of dry stone fill was to pack the stones in the densest state, requiring skilled workers. For instance, in Figure 2.7, the extreme right wing of the dry stone fill was in advance of the upstream masonry and could withstand the H/V=1/2 upstream slope without instability.

The main geology of rock types used were granodiorite, granite, gneiss and schist. The average unconfined compression strength of the rock matrix was 120 MPa; however, less competent rock was used. Detailed investigations identifying and characterizing rock are reported in Table 2.2. The dams investigated by EOF are presented in Table 2.3 and shown in Figure 2.8. Reported minima and maxima of properties were measured recently using samples taken at the quarry. These properties will be compared with the observed performance of the dry stone pitching (see section 2.2.3) and the monitored displacements (see section 2.2.4).

Dam	A	B	C	D	E
Geology type	Granite and Gneiss	Grano-diorite	Grano-diorite	Schist	Grano-diorite
Porosity n (%)	1.8/2.3	1.3/1.5	1.0/1.3	0.2/0.6	0.7/0.8
Continuity index Ic (%)	61/73	63/69	64/74	81/84	72/73
Unconfined strength Rc (MPa)	56/63	135/156	78/117	91/129	153/157
Tensile strength Rt (MPa)	8/11	12/13	10/13	5/6	14/16
Point load test Is_{50}	4/8	10/12	7/10	5/7	13/14
Young's modulus E (GPa)	28/49	44/49	37/52	43/118	47/70
Sulfate of magnesium (%)	0.1/2.1	0.1/0.1	0.0/3.7	0.1/5.1	0.0/0.0
Dry micro deval (%)	3/15	5/6	4/6	4/5	5/5
Wet micro deval (%)	8/35	10/11	7/15	16/22	9/12
Los Angeles (%)	20/37	24/28	19/30	14/14	21/22

Table 2.2. *Rock identification and characterization of investigated rockfill dams (courtesy of EDF)*

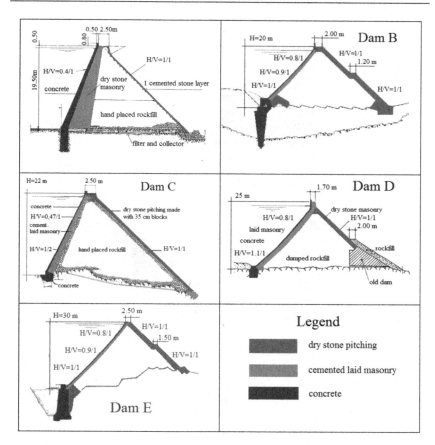

Figure 2.8. *Cross-section of rockfill dams investigated by EDF. For a color version of the figure, see www.iste.co.uk/vincens/drystone.zip*

Dam	A	B	C	D	E
Height (m)	20	20	22.5	25	30
Type	Dumped	Dumped	Hand-placed	Hand-placed	Dumped
H/V(D/S)	1/1	1/1	1/1	1/1	1/1

Table 2.3. *Rockfill dams investigated by EDF*

The gray-colored cells in Table 2.2 highlight some low values that we will discuss later on. From this investigative campaign, it can be concluded that:

– dams B and E are made with competent rock, not susceptible to weathering;

– without a good selection of competent rock by skilled workers, some blocks of dam C are susceptible to climatic influence;

– some blocks of dam D are susceptible to cracking by cycles of extreme temperatures and wetting;

– blocks of dam A are susceptible to cracking or their contact and joints might be weathered over time.

The average size of the stones was around 30 cm. The gradation of the rockfill was poorly graded: the average uniformity coefficient $Cu=d_{60}/d_{10}$ is between 2 and 3. The documents relating to the construction reported that the global porosity of the dumped rockfill was estimated at about 0.4 and the porosity of the hand-placed rockfill at about 0.2.

All the stone pitching placed on the downstream slopes has two layers: a pavement layer and a sub-base layer. All the blocks of the stone pitching lay perpendicular to the slope. They lay on a sub-base layer of smaller stones. The sub-base layer is a more rigid support. It smoothes the surface and avoids local stress concentrations. There are three main types of pavement layers, according to the stone sizing and bonding:

– sized blocks without cement (Figure 2.9(a));

– natural stones without cement (Figure 2.9(b));

– natural stones bonded with cement (Figure 2.9(c)).

2.2.3. *Observed damage on the dry stone pitching*

Four kinds of damage are observed on the pitching stones:

a)

b)

c)

Figure 2.9. *Main types of downstream stone pitching (courtesy of F. Laigle).
For a color version of the figure, see www.iste.co.uk/vincens/drystone.zip*

Figure 2.10. *Crack through a stone (courtesy of F. Laigle)*

Figure 2.11. *Settlements caused by the displacement of cracked stones (courtesy of F. Laigle)*

Figure 2.12. *Dismantling following extruded stones (courtesy of F. Laigle)*

Figure 2.13. *Bulging of the dry pitching stone (courtesy of F. Laigle)*

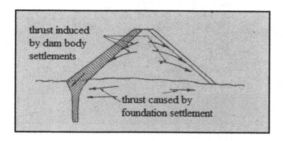

Figure 2.14. *Arching effect in the dry stone rockfill dam [COY 39]*

1) The breaking of stones into two parts occurs sometimes with a bang. It was mentioned in the first report filed for dam D (Figure 2.10).

2) The cracks are not currently repaired. Several cracks in the same zone of the stone pitching lead to local settlements (Figure 2.11) and may have an impact on the global settlement of the dry stone pitching.

3) The extruded stones cause local dismantling of the dry masonry (Figure 2.12) and require maintenance before the dismantling expands to the surrounding area. Coyne [COY 39] underlined an arching effect through the dam body between the upstream and downstream dry masonries (Figure 2.14). It is not clear whether the arching effect through the dam body contributes to the extrusion of some stones. The

addition of compressive and local shear stresses seems to cause the observed buckling of the stone pitching.

4) Several rows of bulging are observed on stone pitching made with rectangular-shaped blocks (Figure 2.13). This kind of dry masonry seems too stiff to adapt to the settlement of the dumped rockfill fill. This last damage requires maintenance works to avoid a general collapse of the stone pitching (Figure 2.15).

Figure 2.15. *Repair of bulges in the dry stone pitching (photo EDF)*

2.2.4. *Dry stone rockfill dam performance*

Five features of deformation behavior have been detected by monitoring. The postconstruction settlement rate per year is presented in Figure 2.16, where the benchmarks and their measured settlement rate in 2001 are attached in a plane view for four of the five dams investigated. The settlements of the four dams, expressed as a percentage of the height, are given in Figure 2.17.

[CLE 84] studied the crest postconstruction settlement of 18 dumped rockfill dams and evaluated the settlement data after 10 years of service. A best-fit analysis resulted in the following relationship:

$$s = a\,H^b \qquad\qquad\qquad [2.1]$$

where:

s = postconstruction settlement (m);

H = height of the dam (m);

a= 0.009   and   b = 0.9.

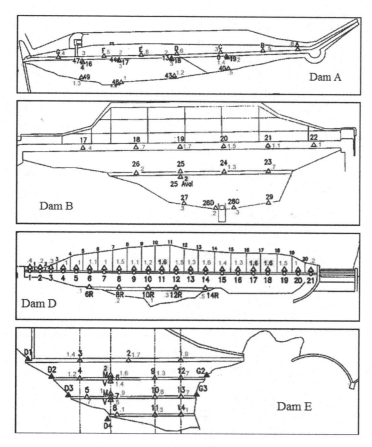

Figure 2.16. *Yearly settlement (mm/year) of dams investigated by EDF*

Assuming that the settlement is proportional to the logarithm of time and the Clements relationship is valid, the crest settlements in 2001 are evaluated at 1.1 mm/y for dams A, B and D and 1.6 mm/y for dam E. However, after more than 50 years of service, the maximum measured settlement rate was found to be higher:

– 1.6 mm/y for dam D;

– 2 mm/y for dams B and E;

– 4 mm/y for dam A.

The second feature is that the secondary consolidation is not complete. Indeed, it is recognized that the secondary consolidation of dams is complete when the settlements expressed as a percentage of the height are less than 0.02%. Figure 2.17 shows that the behavior of dams B, D and E is quite different from that of dam A:

– dam D has an astonishing settlement at its toe, may be not significant.

– dam B has only one benchmark at its crest where the settlement is at the limit of 0.02%.

– dam E has reached the criterion on the first berm only at one point.

– dam A shows a general trend of the downstream face a constant and general trend to settle, with relative settlement larger than 0.02%.

The third feature is that the maximum rates of horizontal displacement at the crest are around 2 mm/y: 1.9 mm/y for dam B, 2 mm/y for dam D and 2.2 mm/y for dam E, which are larger than the maximum settlement rate. Displacement at dam A in contrast is 2.5 mm/y lower than the settlement rate (Figure 2.18).

The fourth feature is the existence of a global movement of the dry stone pitching. Large relative horizontal deformation rates are observed at the downstream toe: 0.0004/year at dams A and B and 0.00024/year at dam E. The measured displacements during 2001, shown in Figure 2.19 by arrows on the downstream face of the cross-section of dam E, illustrate a clear tendency to move parallel to the downstream slope.

The last but not the least feature is the linear trend of displacements versus time. Figure 2.20 shows the settlement of benchmark 44 on downstream face of dam A.

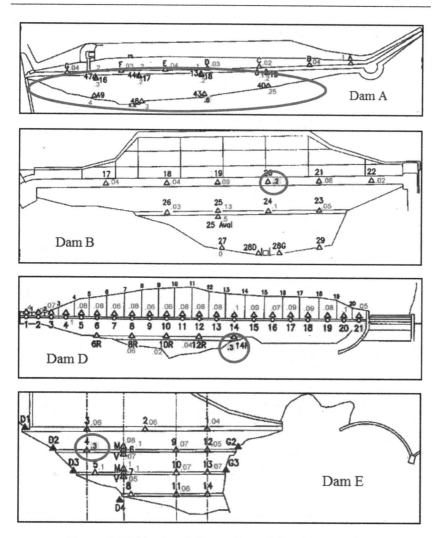

Figure 2.17. *Yearly relative settlement (mm/y) measured
for four dams investigated by EDF. (Red circles show areas where the
threshold 0.0001 mm/year is passed). For a color version of this
figure, see www.iste.co.uk/vincens/drystone.zip*

The main feature of the current hydraulic behavior is the small
leakage measured downstream of the rockfill dams. Leakage is a key
parameter concerning the overall performance of the CFRD: it is an

indication that no opening has occurred at the upstream face or that the concrete face has not cracked. One opening occurred on some dams at the first filling or sometimes a leakage through the soil foundation was detected and was successfully repaired.

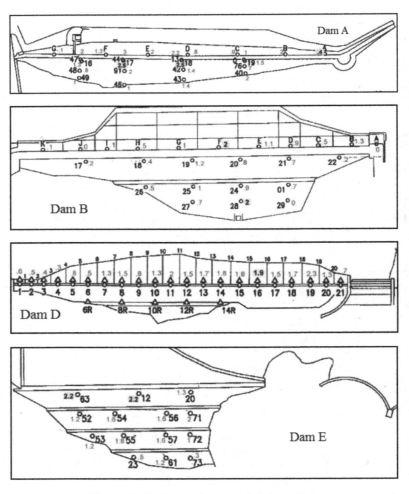

Figure 2.18. *Yearly horizontal deformation (mm/year) of the EDF dams*

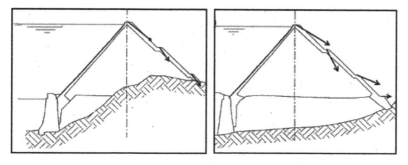

Figure 2.19. *Vector of displacements measured downstream of dam E (courtesy of EDF)*

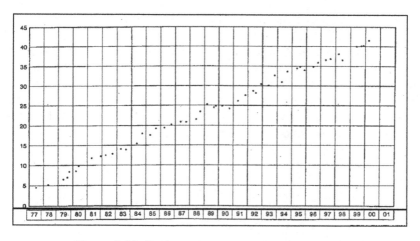

Figure 2.20. *Settlements measured on benchmark 44 of dam E (courtesy of EDF)*

In conclusion, the main features of dam behavior are related to the deformation behavior of the crest and the downstream face. Some permanent residual deformations are monitored and are challenging. They raise two questions:

– What is the ongoing physical process? Is it creep or the initiation of a sliding process?

– What are the consequences of the measured deformations on the dry stone pitching on dam safety?

In view of answering both questions, a short review of the mechanical properties of the rockfill is presented hereafter. The main point studied was the evaluation of the angle of internal friction of the rockfill and the second important point was compressibility related to the residual deformations in the dam.

2.2.5. *Shear strength of rockfill*

The shear strength of soils is generally measured using laboratory triaxial tests. The problem of the rockfill is that the maximum dimension of the blocks is similar to the diameter of the conventional triaxial cells. It is worth remembering that the shear strengths of rockfill are measured on a small fraction of the material.

It is demonstrated that materials exhibit higher shear strength when they have a well-graded grain size distribution, high density, angular shape and dry state. They exhibit a lower shear strength under an increase in the normal pressure on the failure plane. It is confirmed that the behavior of rockfill is nonlinear, and that a relation between the shear stress and the normal stress is of the form:

$$\tau = A * (\sigma')^b \qquad\qquad [2.2]$$

where:

τ = shear stress and σ' = effective normal stress;

A and b = empirical coefficients that depend on the type of rock.

Leps [LEP 70], Barton and Kjaernlsi [BAR 81] and recently Frossard [FRO 05] proposed empirical approaches for the estimate of the block shear strength related to the previous parameters.

2.2.5.1. *Leps' shear strength estimation*

Leps [LEP 70] reviewed the shear strength of compacted rockfill as measured with the use of large diameter laboratory triaxial tests. The data plot, as presented in the paper, is shown in Figure 2.21. The shear strength includes no apparent cohesion and is measured by the angle of internal friction plotted against the normal stress on the failure plane. Friction angles reduce by 6 or 7° for 10 times increase in the normal pressure on the failure plane.

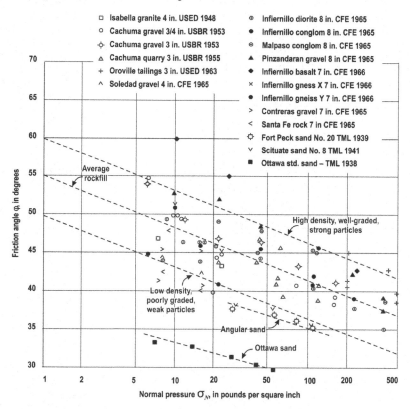

Figure 2.21. *Shear strength of rockfill after [LEP 70]*

Assuming that the normal stress is below 70 kPa (10 Psi) and strong particles are defined by Leps as rocks with an unconfined compressive strength of 69–207 MPa, the angle of internal friction varies from about 45° for low density to as high as 53° for high density, and a very low stress. The application of 50° to the stability of the downstream face of dams A to E gives a safety factor of 1.2. This result is in accordance with the general opinion that H/V=1.3/1 rockfill slope yields a safety factor of 1.5 [ICO 10].

2.2.5.2. *Shear strength estimation from Barton's relationship*

The friction angle is dependent on three state variables: shape of the particle, including microindentations, porosity and effective stress and two strengths, the unconfined compressive strength and the basic friction angle. Its value is estimated from these variables using equation [2.3] and Figure 2.22 [BAR 81]:

$$\varphi' = R \cdot \log(\frac{S}{\sigma_n}) + \varphi_b \qquad [2.3]$$

where:

– R: equivalent roughness, based on the porosity and on the degree of particle roundedness and surface smoothness;

– S: equivalent particle strength, based on the unconfined compressive strength σ_c of the rock and on the D_{50} particle size of the rockfill;

– σ'_n: normal effective stress;

– ϕ_b: basic friction angle, measured on a dry, flat, sawn surface of rock by tilting. It usually ranges from 25 to 35°.

The friction angles of the top layers of rockfill and the dry stone pitching stones are calculated for 70 kPa of normal effective stress by application of equation [2.3] and the values are presented in Table 2.4. These values were calculated with the assumptions that the basic friction angle is equal to 30° for the granite, and 25° for the schist, values used by the authors in their paper. The porosity is at about 40% for dumped rockfill, 30% for hand-placed rockfill and 20% for the dry stone pitching. The unconfined compressive strength is given in Table 2.2.

Dam	A	B	C	D	E
$\phi(°)$ triaxial-$\phi(°)$2D rockfill	43 – 46	44 – 47	47 – 51	44 – 47	45 – 48
$\phi(°)$ triaxial-$\phi(°)$2D pitching	52 – 56	54 – 59	54 – 57	49 – 53	54 – 59

Table 2.4. *Friction angles calculated for dams*
A to E from [BAR 81] at σ_n=70 kPa

The friction angle on site $\phi(D)$ of the full grading $0/D$ with the in place porosity n_D is estimated from the angles $\phi(d)$ and ϕ_b measured in lab on the scalped grading $0/d$ and porosity n_d with equation [2.4].

$$\phi(D) - \phi_b = \frac{R(n_D)}{R(n_d)} \frac{\log_{10}(S_D)}{\log_{10}(S_d)} (\phi(d) - \phi_b)$$ [2.4]

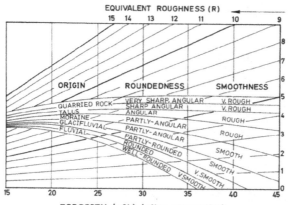

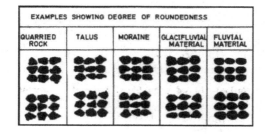

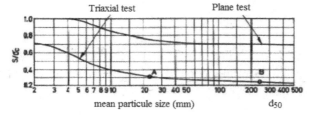

Figure 2.22. *Shear strength of rockfill after [BAR 81]*

2.2.5.3. *Shear strength estimation from Frossard and Bolton's relationship*

The shear strength estimation proposed by Frossard [FRO 05] is based on three equations delivered in the MICROBE research project (equations [2.5]–[2.7]).

$$\tan^2\left(\frac{\pi}{4}+\frac{\varphi}{2}\right)=\left[1-\frac{d\varepsilon_V}{d\varepsilon_1}\right]\tan^2\left(\frac{\pi}{4}+\frac{\psi}{2}\right) \qquad [2.5]$$

$$\psi=\phi_\mu+\theta_M \qquad [2.6]$$

$$\varphi-\psi=A.\left[I_D\left(10-\ln(p')\right)-1\right] \quad \text{(Figure 2.23)} \qquad [2.7]$$

where:

ϕ: peak of friction angle of the rockfill (ϕ_{max} in Figure 2.23);

ψ: critical angle or non-dilative angle of friction (ϕ_{crit});

ϕ_μ: basic friction ratio of a sawn surface of rock;

θ_M: increase in friction ratio caused by the roughness;

$(1-d\varepsilon_V/d\varepsilon_1)$: the dilatancy rate of Rowe. The maximum value is evaluated with equation [2.7] [BOL 86];

p': mean effective stress in kPa;

A: equal to 5 for plane strain test and 3 for triaxial test;

I_D: relative density.

The peak friction angles evaluated with the previous equations [BOL 86, FRO 05] are 42–46° for triaxial test and 46–50° for plane strain condition, assuming that ψ ranges from 36 to 40° and relative density is 50%.

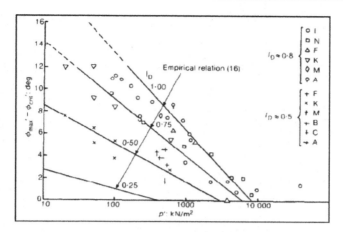

Figure 2.23. *Dilatancy friction (Φ-Ψ) versus effective mean stress [BOL 86]*

2.2.5.4. *Conclusion on shear strength estimations*

The synthesis of the three empirical approaches shows that the internal friction angles at low confining stresses are higher than 45°. It implies that the dams studied have no stability problems as far as the downstream slopes of the rockfill dam behave under plane strain condition.

2.2.6. *Compressibility of rockfill*

The compressibility of the rockfill is highly dependent on the void ratio. Laigle correlated the hyperbolic modulus K_b with the porosity n [2.8].

$$E = K_b \cdot p_a \left(\frac{\sigma_3}{p_a} \right)^m \qquad [2.8]$$

where $K_b = 4/(n)^3$

Assuming that the porosity of the random rockfill is 0.4 and the porosity of the dry stone pitching is 0.2, the modulus of the downstream pitching is 8 times larger than the modulus of the dam body. The void ratio is related to the shape of the blocks and the gradation of the material: Figure 2.24 shows the minimum and maximum void ratios *versus Cu* the coefficient of uniformity and R the block elongation ratio is given by the ratio *d1/d2*, where *d1* and *d2* are the block dimensions in the principal inertia directions [BIA 97].

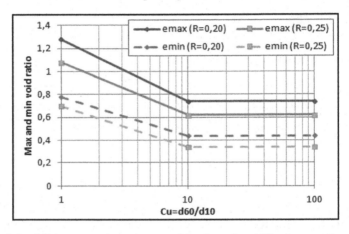

Figure 2.24. *Minimum and maximum void ratios of rockfill versus uniformity coefficient [BIA 97]. For a color version of the figure, see www.iste.co.uk/vincens/drystone.zip*

However, there is a strong mechanical anisotropy associated with the anisotropic shape of the blocks [DEL 04]. The modulus for a loading perpendicular to the largest dimension of the blocks is larger than the one found with a loading parallel to the largest dimension of the blocks. The breakage of the blocks changes the poorly-graded rockfill into a well-graded material (Figure 2.25) and causes a large compressibility. The critical state line identified is then no longer valid and the dilatancy rate is strongly reduced: it is related to the breaking stress σ_b in equation [2.9] [MCD 98].

$$\varphi - \psi = A.\left[I_D.\ln\left(\frac{B\sigma_b}{p'} \right) - 1 \right] \qquad [2.9]$$

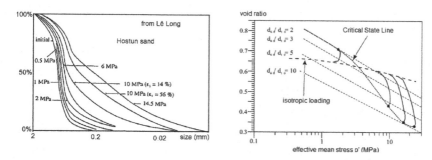

Figure 2.25. *Crushing increases the compressibility of rockfill [BIA 97]*

Hardin's particle breakage factor is defined as $Br = Bp/Bt$ [HAR 85], in which Bt is the total breakage represented by the area between the original gradation curve and final gradation curve, as indicated in Figure 2.26, and Bp=breakage potential represented by the area over the original grain size distribution curve and limited to US sieve No. 200 or $d>74$ μm. It appears that the total breakage is dominated by the total energy input [LAD 10]. The energy input includes effects of frictional sliding and rearrangement of particles, which dominates the behavior at lower confining pressures where very little crushing and small time effects are observed in granular materials with strong particles. However, as the confining pressure, shear stresses and shear strains increase, the particle crushing begins to dominate the behavior. At these higher stresses, the particles are held tightly together, and it is the particle crushing that produces strains rather than the rearrangement and frictional sliding between intact grains.

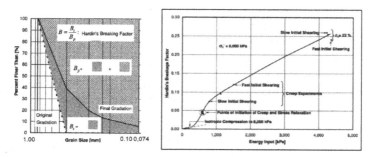

Figure 2.26. *Definition of the particle breakage factor Br [HAR 85] and change of Br with energy input [LAD 10]*

2.2.7. Scale effects

The scale effect has two components of different kinds:

– the mechanical scale effect causes the decrease in the friction angle and the dilatancy rate under increasing stress by the crushing of particles;

– the geometric scale effect is, for instance, the increase in the friction angle and the dilatancy rate by reducing the number of particles.

2.2.7.1. Mechanical scale effect of the crushing

[MAR 67, MAR 73] measured the crushing strength F as the maximum load reached before the collapse of particles placed between two stiff platens. In the range of materials investigated by [MAR 67, MAR 73], the average crushing force of a particle F is a power function of the average size d of the particles with the exponent $\lambda < 2$ (Figure 2.27). Thus, the crushing stress (F/d^2) decreases when the size d increases.

$$F = \eta d^{\lambda} \qquad\qquad [2.10]$$

where η is the crushing strength (kN) of a stone with average size of 1 cm and λ is the exponent found by [MAR 73] to be between 1.2 and 1.7.

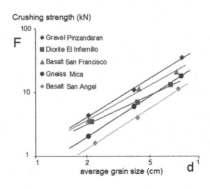

Figure 2.27. Crushing strength versus grain size after [MAR 73]. For a color version of the figure, see www.iste.co.uk/vincens/drystone.zip

Rock materials generally contain a pre-existing distribution of microcracks produced during geological times on igneous rocks during cooling from the magma or on metamorphic rocks exposed to weathering and recently by blasting. These cracks are activated and grow in size due to stress concentration under permanent loading, at particle contacts. The microcracks can coalesce to form larger cracks and can propagate with very little plastic deformation up to the complete fracture of the block [LAD 10]. In the classical theory of fracture mechanics, the intensity of the tensile stress σ propagating a crack of a length l to the complete fracture depends on the toughness K_c and the geometry (coefficient β):

$$\sigma = K_c / \left(\beta . \sqrt{\pi l} \right) \qquad [2.11]$$

In a material, where the crack length l is proportional to the size d of the block, the exponent is $\lambda = 1.5$. In fact, λ is deduced from the statistical distribution of microcracks. The probability P_s that the microcracks do not coalesce and do not fracture the block decreases as the volume V of the block and the applied stress σ increase with the following function distribution, where m is the modulus of Weibull, [WEI 51] which ranges from 3 to 18 with a mean of 6:

$$\ln(P_s) = \frac{V_0}{V} \left(\frac{\sigma_0}{\sigma} \right)^m \qquad [2.12]$$

Exponent λ is derived from m:

$$\lambda = 2 - \frac{3}{m} \qquad [2.13]$$

Furthermore, if a particle has a high number of points of contact with neighbors (a high coordination number), the load on it is well distributed and the probability of fracture is much lower than that at low coordination number. Equation [2.10] was changed into equation [2.14] to include the effect of confining stress σ_m on the crushing load [AUV 75].

$$F = \eta . d^\lambda + b . \sigma_m . d^2 \qquad [2.14]$$

The value of coefficient b depends on the particle shape. It is equal to $\pi/4$ for spherical particles and increases with the coordination number.

2.2.7.2. Geometric scale effect on deformation

The main assumption of the continuum mechanics is that stress and strain fields are homogeneous in a laboratory sample. This is the reason why the maximum particle size is limited in triaxial and oedometer tests. Usually, the ratio of the size L of the tested sample and the maximum size D_{max} of the tested material has to be higher than 10 or 20. It will never fall below 6. Mamba [MAM 89] did some experiments on the fines fraction of the rockfill of Vieux-Pré Dam with two shear boxes. It shows that the influence of the geometric effect extends up to L/D_{max}=50 (Figure 2.28).

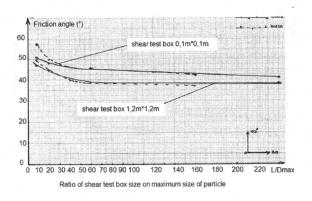

Figure 2.28. *Effect of L/D_{max} on peak friction angle [MAM 89]*

2.2.8. *Time effects*

The time effect during dam operation has three components:

– the cycle-dependent effect: the hardening or fatigue during cycles of the reservoir load causing a part of the monitored creep under frequent water level;

– time-dependent or aging effect due to changes in the environmental conditions which impact on the mechanical properties;

– rate-dependent effect: response of material due to viscous properties.

The response of rockfill material and dams to all these effects is not clear. Only some of these phenomena have been investigated.

2.2.8.1. Hardening effect

The application of energy input is predominantly expended on particle crushing at high stresses. However, at medium stress, the operation of reservoir cycles increases the total energy input applied to the dam. In consequence it increases the total breakage and the dam body compressibility.

2.2.8.2. Aging effect on crushing

Experiments on rock and concrete specimens have clearly shown that their strengths are strongly dependent on time. Figure 2.29 shows the stress–strain–time behavior obtained from tests on concrete cylinders [RUS 60]. As the load on the cylinders is held constant below the short-term fracture load, the time to fracture increases with decreasing load until reaching the residual strength, sometimes called activation load. The fracture does not occur at all under loads smaller than the residual strength. The residual strength the concrete is about 80% of the short-term fracture load.

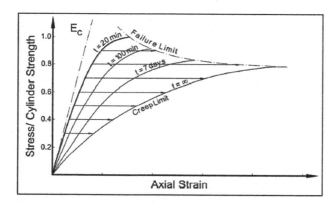

Figure 2.29. *Effect of time on the strength of concrete cylinder after [RUS 60]*

This phenomenon is referred to as static fatigue or delayed fracture [LEM 90]. Single rock blocks behave similarly in the sense that their crushing strengths are time-dependent. The extrapolated values of residual strength of the granite range from 40 to 85% [KRA 80, LAU 99] of the short-term fracture strength. However, the duration of the experiment was not long enough to reduce the uncertainty and a residual value equal to 0 after several million years cannot be excluded.

Experiments on rock and concrete specimens have clearly shown that their strengths are strongly dependent on time. This phenomenon explains the change of compressibility and increasing displacement *versus* time. It means that the friction angle is rate-dependent and creep effect may be dominant in a prefailure state.

It has been observed that the reduction of rock strength is due to Kc [2.14] and to the water. The fractures can propagate at a finite velocity, depending on external "corrosive" agents such as water and degree of humidity. This phenomenon is known as a subcritical propagation of the fracture [CHA 03b]. In consequence, the speed of fracture is strongly influenced by mechanical and environmental factors. Then, the settlement of rockfill dams with time can be understood in terms of stress corrosion and water-enhanced static fatigue of the rockfill [CHA 03b]. [CHA 58] designed a relation between the two tensile strengths σ_1, σ_2 of glass samples and the time to fracture t_1 and t_2 as:

$$\sigma_2 = \sigma_1 \left(\frac{t_1}{t_2} \right)^{\frac{1}{n}} \qquad\qquad [2.15]$$

The particle crushing is a time-dependent phenomenon described as static fatigue or delayed fracture. The deformation of rockfill is a combination of frictional sliding and particle crushing. Frictional sliding is the most important component of deformation at lower stresses, while it plays a much smaller role at high stresses where particle crushing dominates the development of deformations.

2.2.8.3. *Rate effect on strength and crushing*

Only small effects of loading strain rate on stress-strain, volume change and strength behavior were observed, since the close relation between time effects and crushing in granular materials has been established [LAD 10]. However, a higher initial loading strain rate resulted in subsequent higher creep axial strain, higher energy input/volume and higher amounts of grain crushing. In consequence, the high level of strain rate during earthquake will be a larger combination of particle crushing and lower combination of particle sliding than the low level of strain rate provided in the current triaxial test.

2.2.8.4. *Correlation between dam performance and time and climate effects*

The assumption that environment can control the rate of deformations was tested on the five dams studied. The partial factors of correlation of the yearly displacement rates were calculated with five independent variables: time (number of years), reservoir level (annual mean level), frost (number of days with temperature lower than 0°C and number of days with temperature changes higher than +3°C to –3°C, rainfalls (cumulated rainfall height in 1 year) and annual temperature were analyzed. The values of the partial correlation factors are presented in Table 2.5. Time is the main variable explaining the displacement rates so far. It worth noting that one dam is more susceptible to environmental constraints (frost, rainfall and temperature) than the others: dam A. Dam E shows another susceptibility to the environmental constraints, but the nature of this susceptibility is quite different: it is mainly linked to the reservoir operation. The dependency on the climate of the two other dams is not proved (factor of correlation lower than 0.7).

In conclusion, dam A is the most susceptible to the environment. Accordingly, it is the dam with the highest displacement rates and the poorest quality of rock. Melnik [MEL 82] found evidence of weathering in the first 2–4 m of the top layer of the facings when studying several old weak rockfill dams.

Partial factors	Horizontal dam displacements				Vertical dam displacements			
Dams	A	B	D	E	A	B	D	E
Time	0.78	0.94	0.96	0.96	0.83	0.96	0.95	0.94
Reservoir level	0.08	0.04	0.11	0.82	0.15	0.05	0.13	0.78
Frost index	0.30	0.12	0.07	0.09	0.51	0.12	0.09	0.12
Rainfall	0.12	0.00	0.00	0.00	0.05	0.01	0.01	0.00
Temperature	0.07	0.00	0.03	0.01	0.14	0.01	0.01	0.01
RL+FI+R+T	0.73	0.50	0.53	0.90	0.83	0.51	0.49	0.84

Table 2.5. *Partial factors of correlation [TRA 06]*

2.2.9. *Appraisal of the behavior of dry stone rockfill dams*

The conclusions of the overview on the behavior of dry stone rockfill dams are:

– The steep slopes of dry stone rockfill dams are no longer used in modern practice. The state-of-the-art recommends higher safety factors. Nevertheless, no problem of stability has ever been noticed in the fill itself of dams A to E. Problems come from local defects.

– The estimated plane strains friction angle of the random rockfill at low confining stress is higher than 45° from three different approaches and is enough to provide the dam with stability. However, any large heterogeneity or flaw in the dry stone pitching leads to 3D deformations, where the plane strain friction angle is no longer valid and is replaced by the lower triaxial friction angle, triggering instability.

– The visual inspections pointed out the fracture and extrusion of some blocks of the dry stone pitching. Remedial maintenance is required to avoid more dismantling expanding far away from these flaws. For this reason, any flaw of the stone pitching has to be repaired. For example, for dam B, the bulging of the stone pitching is caused by the high rigidity contrast between the very anisotropic dry stone pitching, made up of oriented rectangular blocks (perpendicular to the main stress) and the settling fill made up of random stones.

– The monitored behaviors of dry stone rockfill dams show special features: larger displacements than expected, locally linear trend *versus* time and deformation vectors parallel to the slope.

– The local long-term deformations seem to result from aging or cyclic hardening tendency or both rather than being a sign of shearing in the dam body. They are more closely related to the particles crushing than to the plastic deformation caused by sliding. Dam A shows a tendency to exacerbate the problem; the general deformation trend of the downstream slope of dam A is not only associated with the lowest strengths of the rock and to frost action, but is also due to the compressibility of the soil foundation. For this reason, this dam was reinforced with a downstream berm. The rockfill of dam B in contrast is made of sound rock, however, the aging or cyclic hardening tendency is clearly observed.

– The current margin of safety for this type of dam is not straightforward to evaluate. The main point is that current stability analyses do not take into account the scale and time effects. The main issue is the consequences of the future long-term deformations. In the past, there have been several examples of failure of dry stone masonry, several centuries after completion: for example, the tower of Châlut Castle in 2010 after 8 centuries... Thus, a new method integrating the scale and time effects to predict long-term behavior, such as simulations of displacement rates, could be very useful to operators and owners who want to understand and extrapolate the behavior currently observed.

2.3. Numerical modeling of dry stone rockfill dams

2.3.1. *Introduction*

2.3.1.1. *Objectives and selected method*

The objective of the numerical simulations involved hereafter is to provide information that monitoring could not retrieve. The model was designed to predict the long-term behavior of a dam and to quantify the impact of scale and time effects on its stability.

The interaction between the random rockfill and dry stone pitching is so complex that analytical models are not adapted. However, the finite element method (FEM) is far from its field of application concerning the modeling of dry stone pitching, because the fundamental assumption of continuous mechanics is impossible to respect. Indeed, the problem lies in the ratio between the crest width and the maximum size of the rockfill, which is below 10 and down to 6 (Figure 2.7). It is therefore far from a ratio of 50, required by the continuum mechanics assumption, where stresses can be defined and homogeneous. The use of FEM is far out of its field of application as well, particularly for the crest and dry stone pitching, where elements contain a dozen blocks and are simulated by constitutive laws measured on the lowest fraction of the rockfill (<20 mm).

Moreover, the constitutive equations are not developed to mimic the whole crushable behavior of rockfill. Finally, the DEM seems to be a more suitable method able to capture the scale and time effects. For all of these reasons, a DEM model for the problem of the stability of rockfill dams with dry stone pitching has been developed.

More precisely, the case of rockfill dam B has been addressed and modeled (dam A is not representative of this type of dam: it is made with a weak rockfill and was reinforced). The type of rock used is granite or diorite. The mechanical and physical properties of this rock are good to very good. The modeling has to explain the causes of long-term displacements. Three materials will be modeled: the random rockfill, the hand-placed rocks called dry stone pitching at the downstream side and the lean stone pitching with the concrete slab at the upstream side.

2.3.1.2. *Terms of reference of the rockfill dam modeling*

According to the serious limitation of laboratory-based rockfill shear strength investigations and the inability to test full-scale rockfill samples with gradings and porosities as built, the calibration of the DEM model involved the following stages:

1) Modeling of particle breakage: the set of parameters is selected in view of fitting the relationship between crushing loads and rock block sizes.

2) Modeling of triaxial tests: a 2D biaxial sample containing particles fitting the grain size distribution of the fine fraction of the rockfill is simulated and the set of parameters is calibrated to mimic the stress-strain curves measured in triaxial tests performed by [MAR 67].

3) Modeling of the 2D dam section: it involves the fabric and the laying of particles in the section. The sizes of particles are taken in accordance with the grain size distribution of the material in the actual dam, in such a manner that the scale effect is automatically inserted in the dam stability analysis. Dam stability is usually considered a plane strain problem, which is herein downgraded to a 2D problem for the sake of simplicity before using more sophisticated 3D tools.

4) Simulations of the main situations of loading that dam B has demonstrated: first, construction by applying gravity; second, the first filling; third, the cycles of reservoir emptying and refilling; and finally the seismic loading. For these situations, the model has to manage time and scale effects. The factor of safety related to the stability of the dam has to be finally derived.

2.3.1.3. *Basic assumptions of the numerical modeling*

The DEM is frequently used to represent the mechanical behavior of glass and sand materials. However, unbreakable circular (2D) and spherical (3D) elements are not enough to represent points 1 and 2 of section 2.2.1.2 [MAH 97]. Shape, roughness and crushing have to be included in particle modeling. Two families of DEM are available:

– Contact dynamics, which uses rigid bodies [MOR 94]. Contact forces are generally computed with a Newton method, until a criterion based on the equilibrium of bodies is verified. This method uses an implicit numerical scheme to solve dynamic equations that enables larger time steps than for other methods. The calculation can then be faster in this case. However, the contacts are inelastic as the bodies (or elements) are rigid. It seems that information propagates at an infinite speed in such a model. Hence, the elastic waves caused by seismic solicitation may not be adequately reproduced.

– Smooth DEM [CUN 79] in which elements overlap at contact points. Contact stiffnesses are given in the normal and tangential directions of each contact so that contact forces can be computed. This numerical method is based on an explicit numerical scheme for solving dynamic equations. Small time steps are required in order to keep the system close to equilibrium and then to provide a certain numerical stability to the scheme. The computation step corresponds to a succession of two substeps (Figure 2.30). During the first one, the contact forces are computed from the intensity of the overlap. Then, new positions of elements are deduced from Newton's law [ITA 06]. Here, elasticity is integrated in the constitutive law at points of contact and elastic waves propagate at a finite velocity.

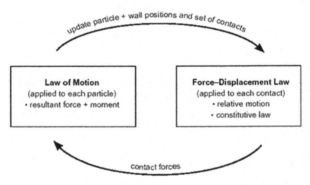

Figure 2.30. *Solving algorithm of a smooth DEM method, after [ITA 06]*

This latter method and the PFC2D software have been chosen here, in order to adequately represent the dynamics of a seismic loading. As the circular or spherical shapes of elements do not match the angularity, shape and roughness of the processed soils and rocks (e.g. the friction angle of spherical or circular triaxial samples does not exceed 25° [MAH 96]), PFC2D was used to model a variety of shapes and angularity of the granular materials. Aggregates of spheres have been developed as in Figure 2.31 [DEL 04] and their suitability to retrieve the high internal friction angles of granular soils has been successfully tested [MAH 97].

Figure 2.31. *Three examples of aggregates of elements [DEL 04]*

An alternative solution to the creation of such complicated shapes was to insert a rolling resistance at contacts between circular [IWA 98] or spherical elements [BEL 09], in order to model the existence of possible plane contacts between actual grains that contribute to increase the shearing resistance of the material, but there was a big disadvantage in this solution: the breakage of particles is associated with the bending moment, and this bending moment in disks is far lower than in real blocks.

In the current work and in previous ones, the first solution has been adopted, because the model has to be based on aggregates of linked elements so that the breakage can mimic the observed crushing during triaxial tests. This is the basic assumption when modeling the time and the scale effects on rock, as shown later on.

Here, displacements of the rockfill dam are analyzed in transversal section. However, the reduction of the 3D geometry to a 2D problem is a simpler solution from a mathematical point of view, but more complex from the physical point of view.

2.3.2. *Constitutive equations and software*

The smooth DEM PFC2D software [ITA 06] has been used to model the aging effect of the dam. This software, developed by Itasca, is suitable for modeling various phenomena such as damage effect in rock mechanics, landslides, powder compaction, and trajectography among others.

The contact force F of a given contact has a normal component F_n and a tangential component F_s:

$$F = F_n + F_s \qquad\qquad [2.16]$$

A linear elastic contact law is generally preferred for its simplicity. In this case, the normal component F_n depends on the normal stiffness of the contact k_n and the normal relative displacement at contact u_n ($u_n = 0$ corresponds to no overlap) following:

$$F_n = k_n u_n \qquad\qquad [2.17]$$

The increment of the tangential force is calculated at each time step as soon as the contact is created:

$$\Delta F_s = k_s \Delta u_s \qquad\qquad [2.18]$$

where k_s and Δu_s represent the tangential stiffness and the relative incremental tangential displacement at contact during the current time step in the contact plane, respectively. Limits are given to the normal and tangential forces, depending on the type of contact used. They will be denoted "simple contact", "contact bond" and "parallel bond" in the following.

The simplest contact incorporates compression force in the normal direction but no tension. As soon as two elements no longer overlap, the corresponding contact vanishes. The tangential force at contact is calculated using equation [2.18], but the value is limited according to Coulomb's criterion where μ denotes the friction coefficient of the contact.

$$F_s < \mu\,F_n \qquad\qquad\qquad\qquad [2.19]$$

Two types of bond between elements can also be used to simulate cohesive behavior. The contact bond corresponds to a special link between two particles at their contact point, which can be broken by tension. It does not prevent rotation between both bonded particles, so shear behavior may not be reproduced correctly. Parallel bonds have therefore been developed and can be broken by a bending moment in the bond. In this case, the bond acts on a given area between the two particles so that rotation between them is prevented. Contact bonds and parallel bonds can be used at the same time, if required. They both have tension and shear strengths and the bond is removed as soon as one of these limit values is reached. Depending on the values given to the limit strengths, it is possible to generate aggregates with either breakable or unbreakable bonds. It is also possible to include these two types of contact in the same aggregate (Figure 2.32). Some particles can then separate but some are not allowed to. In the following, an aggregate denotes an assembly of particles (with breakable and unbreakable bonds), while a clump denotes an unbreakable assembly of circular elements.

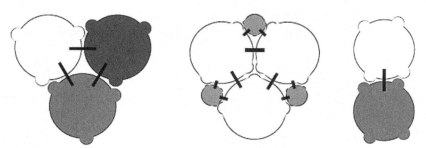

Figure 2.32. *Examples of aggregates of elements constituted of several clumps. Each clump is given a different color from its neighbor [DEL 04]*

Two kinds of damping are also available in PFC2D. The contact damping is applied to the contact and works as a dashpot. Local damping is applied to the sum of forces applied to each particle. The local damping has no physical significance but has a numerical efficiency as it applies not the resultant force acting on each particle, but a certain percentage of this force at each time step. This latter approach is generally used in order to stabilize a system where elastic energy can not vanish or to facilitate the energy dissipation for dynamics problems.

2.3.3. Calibration of parameters

2.3.3.1. The calibration of procedures

Modeling the behavior of a dam requires three main features that are listed below, with their corresponding constitutive laws:

1) The contact between blocks is elastic and frictional. There is no cohesion and no capillarity effect.

2) Rockfill slope should be stable under gravity with H/V=1/1. This corresponds to a friction angle higher than 45° at low confining stress and requires the use of aggregates in order to stabilize the system under gravity.

3) The particles should be able to reproduce the crushing of the blocks and the aging effect [TRA 09]. This is modeled with contact bonds and parallel bonds, whose strength decreases with time. As the limit strength is reached, an aggregate splits into several clumps.

Using the DEM often requires a trial-and-error technique in order to calibrate the local parameters that reproduce the target behavior at a higher scale. This technique was used with a multiscale approach to characterize all the parameters of the model:

– at a small scale, the Marsal tests [MAR 73] have been modeled to calibrate the short-term bond strength acting between clumps. The influence of the bond strength between clumps of an aggregate has also been considered in the long term to take into account the aging effect;

– at a medium scale, short-term biaxial tests have been performed in order to evaluate the influence of the aggregates' shape, the relative density and the local friction angle on the behavior of the rockfill material;

– at a large scale, the calibrated parameters are used to model an embankment under various loading cases.

2.3.3.2. *Small-scale modeling: the crushing load of rock blocks*

First, the normal and tangential stiffnesses were separately calibrated from the other parameters, because they hold mechanical significance as well as influencing numerical features of the computation. In fact, the numerical time step $d\tau$ depends on the ratio $\sqrt{m/k_n}$, where m corresponds to the mass of an individual element. The use of high stiffnesses implies a small time step is required to satisfy correct propagation of the mechanical information through the system. Nevertheless, the calculation time becomes prohibitive as soon as the model involves a large number of elements. Thus, the stiffnesses have been fixed as a compromise between a correct representation of the behavior of the granular soil and a reasonable computation time. A value of 10^8 N/m has been identified for k_n and k_s. This value has also been considered in various studies [CHA 03a, DEL 04, TRA 06, BEL 09, PLA 10]. Once the values of stiffnesses have been established, they are constant during the rest of the study. This separated calibration is possible because the elastic parameters have a limited influence on the strength [BEL 09, CAL 03]. For a faster convergence toward a stabilized state, local damping is also used for all the quasi-static simulations. A damping coefficient value of 0.7, recommended by [ITA 06], has been chosen. A sensibility analysis was carried out on the value of the local damping and negligible influence was observed (<1%) on the soil deformation response under low loading (<0.05 m/s). However, a clear influence of the damping value (>10%) was demonstrated under seismic loading for deformation rates higher than 0.2 m/s.

Then, the Marsal test was modeled to characterize the bond strength of a crushable material. More precisely, the Marsal test characterizes the dependency of the crushing load F_{rupt}^{3D} to the size d of the rock blocks (Figure 2.33). Equation [2.10] can be written as:

$$F_{fail}^{3D} = \eta^{3D} \left(\frac{d}{d_1} \right)^{\lambda^{3D}}$$

[2.20]

where η^{3D} and λ^{3D} are the unit strength and the exponent measured in Marsal tests. In order to use standard units, the unit system (kg and cm) used by Marsal has been changed and replaced by the IS unit. The unit for length d is the meter (m) instead of the centimeter (cm), with $d_1=1$ m, and the unit for strengths F and η is the Newton (N) instead of the kg.

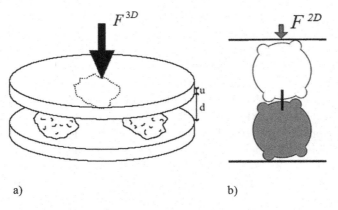

a)   b)

Figure 2.33. *Marsal test: a) 3D experimental set up and b) the corresponding numerical 2D model [DEL 04]*

This experiment has been adapted for 2D simulations [DEL 04]. The assessment of the crushing load distribution in 2D, without experimental evidence, is a complicated and widely discussed issue and has required several numerical checks. First, the analysis of the contact force distribution in 2D and 3D numerical samples composed of clusters simulating blocks has shown that the normalized distribution of contact forces is very similar in 2D and 3D. Second, in order for a 2D computation to represent quantitative features observed in 3D, the probability of particles of size d fracturing has to be the same in 2D as in 3D. The average contact force for each particle size is expressed by Marsal [MAR 73]:

$$F^{3D} = K_M^{3D}\sigma.d^2 \qquad\qquad [2.21]$$

where K_M is a coefficient that depends on the shape of the particles, on the number of particles of the same size d in the granular medium and on the whole grading curve of the material. For a given rockfill material, it can easily be measured from test data. K_M was calculated for a numerical sample and was logically found to be proportional to the rock size d [DEL 04]. Thus, for a 2D numerical sample, the average contact force for particles of size d is given by:

$$F^{2D} = K_M^{2D}\sigma.d \qquad\qquad [2.22]$$

Assuming that the probability of breakage of a particle of size d in the 2D numerical model must be the same as in the 3D real material, and assuming that the contact forces for each particle size follow a normal distribution with the same relative variance in 2D and 3D, the ratio between the average contact force and the average crushing load must be constant in 2D and 3D. The clump crushing load F_{fail}^{2D} with η^{2D} the unit strength for $d=d_1=1$m and λ^{2D} the exponent is defined by:

$$F_{fail}^{2D}=\eta^{2D}\left(\frac{d}{d_1}\right)^{\lambda^{2D}} \qquad\qquad [2.23]$$

$$\frac{\overline{F}^{3D}}{\overline{F}_{rupt}^{3D}} = \frac{\overline{F}^{2D}}{\overline{F}_{rupt}^{2D}} \qquad\qquad [2.24]$$

$$\eta^{2D} = \frac{K_M^{2D}}{K_M^{3D}}\eta^{3D} \text{ and } \lambda^{2D} = \lambda^{3D}-1 \qquad\qquad [2.25]$$

The values of these parameters were calibrated in the breakage particle test (Figure 2.33) performed by [MAR 67]. Two rockfills tested by [MAR 67] are similar to the grano-diorite of dam B: the diorite of Infernillo Dam and the gneiss from Mica Dam (Figure 2.34). The main properties of dam B rockfill are D_{50}=15 cm, the block elongation ratio is estimated at around R=0.3, the uniformity coefficient is between 2 and 3. Finally the initial porosity is estimated at n = 0,45, with a relative density of I_D =60%. The friction angle is assessed using Barton in Table 2.4 [BAR 81].

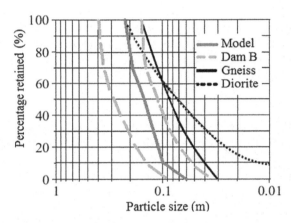

Figure 2.34. *Gradings of the model, dam B, Infernillo Dam diorite and Mica Dam gneiss. For a color version of the figure, see www.iste.co.uk/vincens/drystone.zip*

The compression strength of this material is characterized by the parameters η^{3D} = 3.41.10^6 N, λ^{3D} = 1.6 and K_M^{3D} = 2.69. The 2D parameter K_M^{2D} is evaluated to 0.7 [DEL 04], so that the relationships above give η^{2D} = 8.87E^5 N, λ^{2D} = 0.6. As the assumption of the value K_M^{3D} = 2.69 presents a large part of the uncertainty, a second set of parameters (η^{2D}, λ^{2D}) has been defined in order to seize the influence of K_M^{3D}. The first set is named weak rockfill, while the second set is named sound rockfill in Table 2.6.

	η^{2D} (N)	λ^{2D} (-)
Normal rockfill	$8.87\ 10^5$	0.6
Sound rockfill	$3.45\ 10^6$	0.6

Table 2.6. *Marsal's parameters for the normal and sound rockfills [MAR 67]*

The numerical model of the Marsal test (Figure 2.33(b)) is composed of two plates and an aggregate, representing a rock assembly between them. The plates are moved at a given velocity ($V=10^{-6}$ ms^{-1}) that has to be quasi-static so that the force measured is not influenced by dynamic effects. The aggregate is fixed to the plates so that it cannot topple over during the test. The geometrical anisotropy of the assembly induces a strong variability in compression strength according to the bond orientation (Figure 2.34). Furthermore, the experimental breaking of rocks is ruled by the tension and the shear forces acting on it. Here, the numerical test configuration does not generate tension force on the bond regardless of the aggregate orientation. This test requests shear force and compression force on the bond, for which no failure in compression is allowed. As a result, the strength depends on the shear force only and varies strongly with the orientation of the bond. Thus, this test is repeated with aggregate orientations varying by a step of 10° with respect to the compression direction from one test to the next. Each of the corresponding compression strengths is used to design the mean value $\overline{F_{fail}^{2D}}$, except for non-representative values such as the higher one in Figure 2.34.

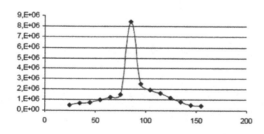

Figure 2.34. *Compressive strength of the numerical block depending on the aggregate orientation*

A parametric study has been required to establish the relationship between the local stength of the bond F_{fail}^{local} and the strength of the assembly F_{fail}^{2D}. It was proposed by [DEL 04].

$$F_{fail}^{local} = \kappa d^{\mu} \left(F_{fail}^{2D} \right)^{\nu}$$ [2.26]

K, μ and ν are the coefficients that have to be defined with a trial-and-error method. A calibration method that has been proposed by [DEL 04] and [TRA 06] was used in the current work. The corresponding results are given in Table 2.7 and Figure 2.35. Moreover, the strength dependency on the particle size is presented for the weak and sound rockfills. The resulting strengths are very different for normal and sound rockfills. This brings out the influence of parameter K_M^{3D}, and the importance of evaluating its value for the *in-situ* relative density I_D.

	κ	μ	ν
Average rockfill	2,850	0.29	0.52
Sound rockfill	10,000	0.4	0.47

Table 2.7. *Parameters sets* (κ, μ, ν) *for the weak and sound rockfills*

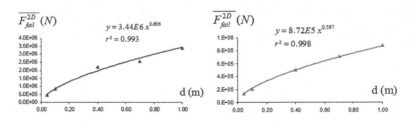

Figure 2.35. *Dependency of strength on aggregate size for sound and weak rockfills*

2.3.3.3. *Small-scale modeling: time effect on the crushing load*

The long-term behavior of rock aging effects is taken into consideration hereafter. The model introduced here has been developed by [TRA 06]. Modeling creep effect implies considering physical time, so that the failure of a bond under constant loading could occur. Let F_{app}^{N} and F_{app}^{S} be the normal and tangential forces acting on a breakable bond in the current step. Let F_{fail}^{N} and F_{fail}^{S} be the normal and tangential strengths of the same bond at the current time step. Failure of the bond occurs as soon as one of the strengths is reached by the activation value. Finally, creep is considered active if one of the values F_{activ}^{N} and F_{activ}^{S} is exceeded at a given time step (Figure 2.36).

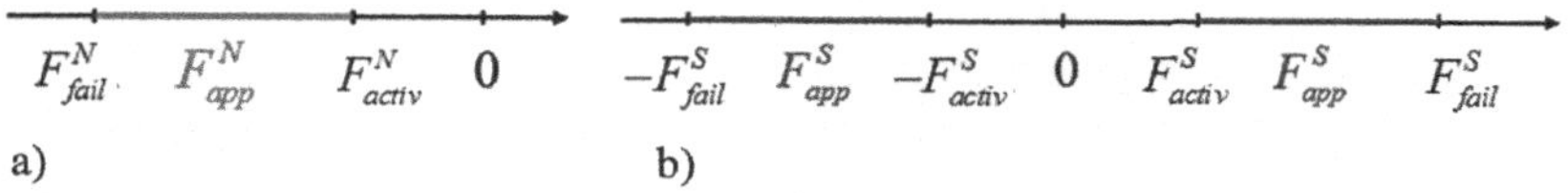

a) F_{fail}^{N} F_{app}^{N} F_{activ}^{N} 0

b) $-F_{fail}^{S}$ F_{app}^{S} $-F_{activ}^{S}$ 0 F_{activ}^{S} F_{app}^{S} F_{fail}^{S}

Figure 2.36. *Creep criteria for the normal force a) and for the tangential force b)*

The aging model has been proposed by Itasca. It is assumed to be similar in both normal and tangential directions. Let F_{fail}^{T} and $F_{fail}^{T+\Delta T}$ be the strengths at two consecutive time steps T and $T + \Delta T$. The value of $F_{fail}^{T+\Delta T}$ is deduced from F_{fail}^{T} with the next relationships where β_1, β_2 and β_3 are the three coefficients to calibrate:

$$F_{rupt}^{T+\Delta T} = F_{rupt}^{T} - \beta_3 \int_{T}^{T+\Delta T} \chi dt \qquad [2.27]$$

$$\chi = \exp\left(-\beta_2 \frac{\left(F_{app}^{T} - F_{activ}\right)}{F_{rupt}^{T}}\right) \qquad [2.28]$$

The model mimics three basic phenomenon: (1) it allows the damage to be added even for different forces which can be successively applied; (2) the strength ranges from the maximum value at quick strain rate to the minimum one, named the activation value;

and (3) the rate of strength drop decreases as the gap between the load and the activation strength decreases.

$$F_{active} = \beta_1 F_{fail}^{T_0} \qquad [2.29]$$

The activation force is controlled by the initial strength $F_{fail}^{T_0}$, where T_0 corresponds to the initial time and β_1 is another constant coefficient.

Assuming that $F_{app}^{T+\Delta T}$ is constant during the time interval ΔT, equation [2.27] can be rewritten:

$$F_{fail}^{T+\Delta T} = F_{fail}^{T}\left(1 - \beta_3 \chi \Delta T\right) \qquad [2.30]$$

The failure of the bond occurs during the time interval ΔT, if the condition $F_{fail}^{T+\Delta T} > F_{app}^{T}$ is no longer verified. Indeed, the duration when breaking occurs ΔT_{fail} under the load F_{app} and the maximum lifetime, T_{max}, for $F_{app}=F_{activ}$, is given by:

$$\Delta T_{fail} = \frac{\left(1 - \dfrac{F_{app}^{T}}{F_{fail}^{T}}\right)}{\beta_3 \chi} \qquad [2.31]$$

$$T_{max} = \frac{1 - \beta_1}{\beta_3} \qquad [2.32]$$

This model has been applied to model the creep effect of several rock samples [TRA 06]. It is able to quite confidently reproduce experiments involving several granite types reported in [SCH 78], [KRA 80] and [SCH 78]. The model is quite efficient at simulating the complete weathering of rock with the simplification $\beta_1=0$. In this case, the creep phenomenon becomes active even if no loading force acts on it. Here, equation [2.32] simplifies and the maximal lifetime of an unloaded bond is given by:

$$T_{max}=1/\beta_3 \qquad\qquad [2.33]$$

In view of analyzing the impact of a rather quick breakage or considerable weathering, parameters β_2 and β_3 have been calibrated considering the next two assumptions under constant loading:

– the bond fails after 10 years if the loading force is equal to 75% of the short-term strength of the bond;

– the bond fails after 100 years if the loading force is equal to 50% of the short-term strength of the bond.

For the given short-term strength, the parameters have been evaluated at $\beta_2=10$ and $\beta_3=6.7.10^6$ N, the strength reduction ratio *versus* time is presented in Figure 2.37 for both loading cases.

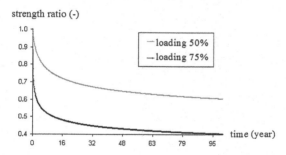

Figure 2.37. *Change in strength over time for a loading equal to 50 and 75% of the short-term strength*

2.3.3.4. *Medium-scale modeling: calibration on triaxial test*

Once the bond strength is characterized for the different sizes of blocks, the grain size distribution is modeled in the representative elementary volume (REV). The short-term behavior of an REV is modeled using biaxial tests. The behavior of the sample depends on the bond strength as well as on the roughness (geometry and friction coefficient) of the elements (clusters) and the porosity between elements. As the roughness and the bond strength have been fixed, the friction coefficient and the porosity become the only parameters to adjust.

Biaxial test simulation requires several phases for the sample preparation. First, 1,500 aggregates are generated in a box that is made up of four plates which serve to apply the isotropic stress (Figure 2.38).

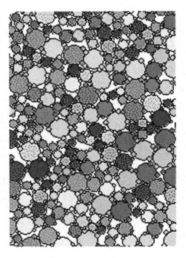

Figure 2.38. *Aggregates forming an REV for numerical biaxial tests. For a color version of the figure see www.iste.co.uk/vincens/drystone.zip*

The physical properties, the model parameters for the material involved in the sample and the confinement for the tests are listed in Table 2.8. The void ratio corresponds to that of the rockfill dam presented later. The distribution of aggregate size conforms to the experimental particle size distribution, but with a certain reduction coefficient. The grading is then shifted to smaller diameters compared to the actual one. In order to generate a sample with a given void ratio, the friction coefficient at particle contact is temporarily decreased. In a third phase, the friction coefficient is reset to its initial value. Finally, the sample is sheared by the deviatoric stress until failure. Figure 2.39(a) shows the evolution of the q/p ratio, where p and q are the mean and the deviatoric (second invariant of the stress tensor) stresses. The corresponding internal friction angle ϕ is deduced from the maximum value reached by q/p:

$$M = \max\left(\frac{q}{p}\right) = \frac{6\sin\varphi}{3-\sin\varphi} \qquad [2.34]$$

Physical characteristics	Value	Mechanical parameters	Values
$Cu = d_{60}/d_{10}$	2	σ_2 (MPa)	0.2 ; 0.5 ; 1.0 ; 1.5
$I_D = (e_{max}-e)/(e_{max}-e_{min})$	60	k_N (N/m)	10^8
n_{Agr}	1500	k_S (N/m)	10^8
		$\tan\psi$ (-)	1.5
		η_{2D} (N/m$^{-\lambda}$)	8.87^{E5} / 3.45^{E6}
		λ_{2D} (-)	0.6

Table 2.8. a) Physical parameters and b) mechanical parameters for the biaxial test

Figure 2.39(b) indicates how the values of ϕ decrease with higher confining stresses. This is in agreement with weak and strong particle rockfill in [LEP 70]. Furthermore, numerical results are in good accordance with the experiment, as the friction angle of the granite is included in the range of the numerical weak and strong materials (Figure 2.39(b)).

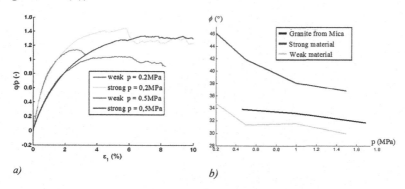

a)   *b)*

Figure 2.39. a) Biaxial tests on weak and strong material for two confining pressures; b) decrease in the internal friction angle with the confining pressure. For a color version of the figure see www.iste.co.uk/vincens/drystone.zip

2.3.4. *Large-scale modeling: construction of the analyzed section*

In this section, we present the different steps for the study of the behavior of a rockfill dam with a dry stone pitching including the creation of the dam section, the impounding and finally the effect of aging on the system. The results of the analysis are giving in section 2.4

2.3.4.1. *Dam section construction*

The previous calibration has been applied to the modeled section of a rockfill dam with a dry stone pitching (Figure 2.40). This corresponds to a simplified version of dam B in Figure 2.8. Indeed, the upstream face of dam B presents a stiffer slope at crest (H/V=0,8/1) than at toe (H/V=1/1,1), while a straight slope (H/V=1/1) has been considered in Figure 2.40. This simplification is quite necessary in order not to make the model creation too complex.

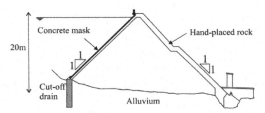

Figure 2.40. *Section of the modeled dam. For a color version of the figure see www.iste.co.uk/vincens/drystone.zip*

The geometry is complex and each part of the dam has to be modeled after an adapted process [DEL 04]. The rock foundation of the dam is modeled with overlapping circular elements to provide a sufficient roughness to the contact with the dam and avoid a slippage which would not be realistic (Figure 2.42(b)). The steep slopes of the upstream and downstream facings, as well as the inclined basis of the dam, do not allow generation of the aggregate the same way as the one used for samples of biaxial tests. In the following, all cells located close to the dam outline are called "particular cells" (Figure 2.41(b)), while the other ones, which have a rectangular shape, are named "regular cells" (Figure 2.41(a)). The aggregates of one regular cell are generated the same way used for the samples of biaxial simulations. Then, all regular cells of the modeled dam are created by duplication from this initial cell. Particular cells are generated with an adapted method, due to their particular shape.

Plates apply a given stress, while an artificial gravity acts in the normal direction to the facing (Figure 2.42(a)). This method is not required for the aggregate in contact with the basis of the dam, because roughness is required in order to prevent any artificial slip surface. Let us consider a regular cell that intersects the dam basis. Aggregates crossing the dam geometry and the basis are embedded with the elements belonging to the rocky foundation (Figure 2.42(b)). Aggregates located downward from the basis are simply deleted. The other aggregates contribute to the rockfill dam.

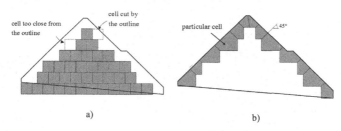

Figure 2.41. Model of the dam; in gray, a) the
regular cells and b) the particular cells

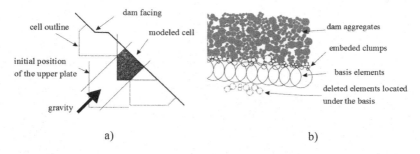

Figure 2.42. Model of the dam; a) generation of particular cells under inclined
gravity loading and b) treatment of the aggregates intersecting the basis

Finally, the concrete face and the dry stone pitching placed on a sub-base are modeled like one and two strips, respectively, of blocks made up of two elements initially bonded together. The blocks are bonded together in the concrete face and placed on one another in the dry stone pitching in order to give it a flexural capacity. Tipping points of the rip rap involve fewer contact points and may exhibit inappropriate flexural ability (Figure 2.43). They have been modeled with bonded blocks, to prevent large deformations of the pitching during construction (Figure 2.43). The other contacts received the same normal and tangential stiffnesses as those used for the aggregates. Finally, the resulting model involves around 20,000 elements.

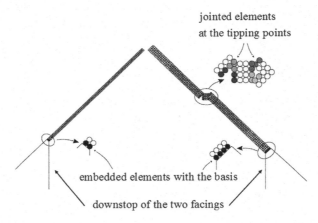

Figure 2.43. *Model for the concrete mask and the rip rap*

2.3.4.2. *Impounding*

The impoundment is modeled with one punctual force acting on each block of the concrete face, corresponding to the sum of the hydrostatic pressure on the upstream faces of the elements. The loading is incremental with the increase in the water height from a hundredth of the final height. Between two increments of loading, the system is allowed to find the equilibrium. Finally, the loading involves 100 stabilization phases of 10,000 cycles, i.e. one million cycles and more are required for the final static equilibrium of the system to be established.

2.3.4.3. *Creep analysis during the dam operation*

Let us consider the aging process at REV and dam scale. We must recall that the acting force differs from one bond to another. At the current time T, the smaller time duration ΔT_{min} corresponding to the next bond failure is calculated by equation [2.31] (in practice, to save computation time, ΔT_{min} corresponded to the next five or 10 bond failures). During ΔT_{min}, convergence is looked for up to equilibrium after applying the loading steps (filling or emptying the reservoir). The corresponding solving algorithm is described below. Assuming that an equilibrium state has been reached at time *T*, taking the aging effect into consideration consists of the following steps:

– determination of ΔT_{min} among the bonds on which creep effect acts. The value of ΔT_{min} is deduced from [2.31]. Each bond corresponds to a given value of ΔT_{fail}. ΔT_{min} corresponds to the minimum value of all values ΔT_{fail};

– determination of the new strength of each contact for the time $T + \Delta T_{min}$. The bond that rules ΔT_{min} is removed;

– physical time is increased: $T \rightarrow T + \Delta T_{min}$;

– search for a new equilibrium state for the time $T + \Delta T_{min}$. Here, creep is not considered because this time increment corresponds to a small physical duration compared to the duration of creep effect. Particle breakage is possible during this phase, only if the force acting on the bond reaches the failure strength.

This last step can have two outcomes. If a new equilibrium state is verified, a new aging effect is considered. However, it is also possible that no equilibrium is reachable anymore. In this case, the failure of the model (sample, dam) occurs and the simulation ends.

Up to now, a model of the short-term and long-term behavior of rockfill material has been developed and calibrated. In the following, these results have been applied to model the behavior of a rockfill dam.

2.4. Results of analysis and interpretation

2.4.1. Stability analysis

2.4.1.1. Numerical criteria to quantify the dam stability

The assessment of the stability from a numerical point of view is not simple. The stabilization involves a very long process of breakages and is very time-consuming. In fact, crushing causes creep and induces local instabilities requiring a very large number of cycles (up to several million) before a convergence. Thus, several criteria were tested. Finally, the dam stability is assessed when two criteria are fulfilled at small and large scales. At the global scale, the dam is considered stable if no significant sliding of entire areas is noticed, or if the change of normalized potential energy is negligible, i.e. the ratio of potential energy by the total mass of the dam drops below the threshold of 10^{-3} m^2/s^3. At the local level, the dam is considered stable, if no significant movement of blocks at the surface of the dam is noticed. This is assessed by considering that the average of the 100 highest velocities of the dam blocks is negligible. Thus, the local criterion is reached when the ratio between the previous normalized velocity and the velocity of the largest block in the dam falling freely from its height drops below 10^{-2} [DEL 04].

2.4.1.2. Definitions of the factor of safety

The stability criteria discriminate stable from unstable states. Now, the margin of safety is usually quantified by the factor of safety. The current definition of the factor of safety is the ratio of the resistance of the system to the driven loading. Herein, it is computed by the maximum factor that reduces the strength of materials up to the limit equilibrium. Partial safety factors can also be defined for each material. Two partial factors of safety were considered in the discrete numerical model of the rockfill dam: the first one for the inner dumped rockfill, and the second one for the outer dry stone pitching. The safety factor for the dumped rockfill, F_R, is defined in [2.35], by considering a reduced internal angle of friction ϕ^* in the model. The safety factor for the dry stone pitching, F_P, is determined by [2.36], by considering a reduced friction ratio, μ^* between the rocky blocks defined in the dam model. The main advantage of this definition of the factor of safety and the use of the reduced local friction ratio is the use of the same initial model.

$$F_R = \frac{\tan\phi}{\tan\phi*}$$

[2.35]

$$F_P = \frac{\mu}{\mu*} = \frac{\tan\psi}{\tan\psi*}$$

[2.36]

Another way to compute the factor of safety is to find the maximum factor that increases the load up to the limit equilibrium. Two additional safety factors have been considered. The first one is for the mechanical loading and the second one for the hydrostatic loading. As the loading of the dam is mainly linked to gravity, the load is increased, considering a rotation of gravity, defined by an angle γ. Then, for an initial slope having an angle β with the horizontal, the safety factor is defined as the ratio between the tangent of the ultimate slope ($\beta + \gamma$) and the tangent of the initial slope.

$$F_\gamma = \frac{F(g)}{F(g*)} = \frac{\tan(\beta+\gamma)}{\tan\beta}$$

[2.37]

The second one is for the hydrostatic loading. The safety factor F_{hw} is defined as the maximum factor that multiplies the water pressure in such a manner that the dam reaches the limit equilibrium.

2.4.1.3. *Factors of safety at the end of construction*

The numerical simulations were carried out with the block shapes of Figure 2.33(b) designed with parallel bonds and a local friction ratio of 0.7 [DEL 04]. The sizes of the blocks range from 7 to 20 cm. The dam model was generated with the weakest rockfill in a very loose state (e = 0.31 in 2D), with a low friction angle of 37° under a confining stress of 100 kPa, giving the minimum safety factor of such a dam section (Figure 2.21).

Without a dry stone pitching, the rockfill dam cannot be stable. The rockfill safety factor is in fact equal to F_R=0.8. The role of the dry stone pitching, which stabilizes the downstream and upstream faces of the dam, with a rockfill safety factor of F_R=1.2 is noteworthy. This value is consistent with the opinion of practitioners.

2.4.1.4. *Factors of safety at first filling*

The stability of the dam with a full reservoir was first analyzed by the rotation of the gravity, as mentioned previously. Only two rotations were tested: the dam was stable after rotating 5°, and unstable after rotating 10°. This means that the mechanical loading safety factor F_γ ranges from 1.19 to 1.42.

Second, the stability was analyzed by reducing the friction coefficient from 0.7 to 0.5 during the first filling, corresponding to F_R=1.4. Displacements presented in Figure 2.44 are large, but convergence was reached after more than 300,000 simulation cycles. The largest deformations occurred in the dry stone pitching just below the crest. As the friction coefficient decreases, the rockfill aggregates become less stable, but their stability is provided by the confinement of the dry stone pitching. A reduction from $\mu = 0.7$ to $\mu^* = 0.1$ was tested, confirming that no equilibrium was reached and the dam collapsed.

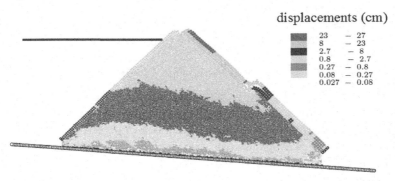

displacements (cm)

23	– 27
8	– 23
2.7	– 8
0.8	– 2.7
0.27	– 0.8
0.08	– 0.27
0.027	– 0.08

Figure 2.44. *Deformations after reducing μ= 0.7 to μ^*=0.5. For a color version of the figure, see www.iste.co.uk/vincens/drystone.zip*

2.4.2. Influence of breakages on dam deformations

2.4.2.1. Analysis of the dam deformations caused by the first filling

At the end of the first filling, the maximal computed displacement is 11.6 cm and is the central deflection of the upstream face of the dam (Figure 2.45). The effect of reservoir loading consists mainly of a horizontal shear of the dam body. The deformations are mainly horizontal, as the maximum settlement does not exceed 2 cm. In this figure, a crack could be suspected at the upstream toe. In Figure 2.46, we can note that about 60% of the deformation occurs in the last 15% of the reservoir filling. Moreover, a tilting of the crest can be observed at the end of the filling process.

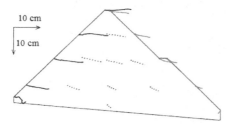

Figure 2.45. *Vectors of displacements at given points of the dam and caused by the first filling. For a color version of the figure, see www.iste.co.uk/vincens/drystone.zip*

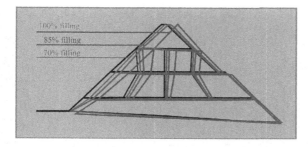

Figure 2.46. *Deformed shapes caused by 70, 85 and 100% first filling. For a color version of the figure, see www.iste.co.uk/vincens/drystone.zip*

The displacements during the first filling are unknown. However, the maximum displacements are estimated by analogy with monitored displacements on dam E at 14 cm in the horizontal direction and 7 cm in the vertical direction. The computed displacements are quite similar to the estimated ones in the horizontal direction and half in the vertical direction. During the first filling, the upstream toe of the concrete broke; this is confirmed by the deformed shape at the end of the filling in Figure 2.46, where the upstream toe moved downstream.

2.4.2.2. *Analysis of the dam deformations after one year and half of service*

One year and a half of service were simulated, with the first filling, the first emptying and a second filling [DEL 04]. The recoverable deformation was found to be quite large on the upstream face, in the order of 10 cm, confirming that the contact stiffnesses were too low (Figure 2.47). However, permanent displacements are quite interesting to observe. The permanent deformation of the downstream face is mainly horizontal. Its maximum value is found on the berm: it is caused by the bulging of the dry stone pitching. The upstream face is lifted by the reservoir filling and settled during the emptying. The crest settled by several centimeters during the first filling, but settled 5 to 6 times more during the emptying. The deformations of the second filling are quite similar to the ones obtained throughout the first filling, though slightly smaller. We can conclude that the cycles of operation contribute to a large extent to the permanent displacements of the dam body.

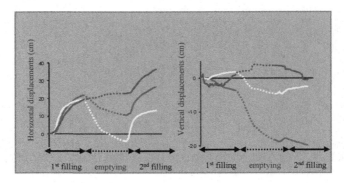

Figure 2.47. *Displacements of three benchmarks caused by 1 year of operation. For a color version of the figure, see www.iste.co.uk/vincens/drystone.zip*

2.4.2.3. *Analysis of the impact of crushing after construction and filling*

A sensitivity analysis was made on the strength of the blocks to check the effect of crushing on the deformation field. The dam model is first built considering unbreakable aggregates (clumps) and loaded by the reservoir. The point is to capture the influence of compaction without crushing. The computed deformations caused by compaction are quite large; the maximum settlement reaches 30 cm at crest. They distort the initial shape. No judgment can be given about that distortion: on site, the rockfill was placed layer by layer and the deformations were not monitored and cannot be compared to the computed ones. In the model, the dam was placed suddenly in a single stage, so both the final fields of deformations are expected to be quite different. For instance, on site, at the end of completion, the displacement is equal to 0 at crest while, in the model, the displacement is maximum at crest. Finally, we have to keep in mind that these large deformations can be associated with the larger porosity of the rockfill assumed in the model (Figure 2.48).

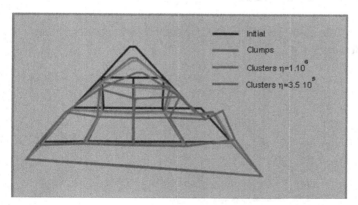

Figure 2.48. *Deformed section caused by compaction and crushing. For a color version of the figure, see www.iste.co.uk/vincens/drystone.zip*

The objective to obtain equilibrium with clumps is the shortening of the CPU time required by convergence. Once the dam has been stabilized, the clumps are changed into clusters. This operation induces a change of a number of elastic contacts (a cluster is made with several clumps and the contacts between the clumps of the cluster are no more rigid and behave with elastic stiffness). In consequence, the change of elasticity triggers a different global deformation field for the dam body. The order of magnitude of this numerical bias is in the order of 5 cm for the settlements at crest.

From this initial state, more realistic deformations are obtained when the crushing of the blocks is triggered, by changing the strength of the blocks. With an exponent λ of 0.6 and the unit block strength $\eta=1.10^6$ N no block breakage takes place and the maximum settlement at crest compared to the clump case remains negligible. If the exponent is kept constant and the unit strength reduced to $\eta=7.5.10^5$N, only one breakage is observed, and the subsequent maximum settlement at crest is found at about 1 cm.

By decreasing η down to 5.10^5N, 200 blocks experience breakage which is about 1% of the total number of blocks (Figure 2.49). In this case, at crest, the settlements increase to 6.5 cm. At downstream face, the bulging of the dry stone pitching is clearly observed. At the base of the dam, the percentage of broken blocks is higher than in the other parts of the dam due to the existence of large stress values acting on the clusters.

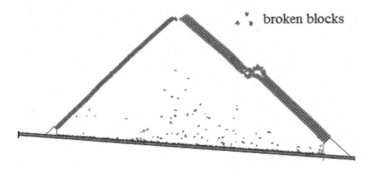

broken blocks

Figure 2.49. *Deformed section and broken blocks with $\eta=5.105N$ after construction. For a color version of the figure, see www.iste.co.uk/vincens/drystone.zip*

With $\eta=3.5.10^5$ N, breakages reach 12% of the total number of blocks. Figures 2.50(a–b) show, respectively, the isocontours of displacements and the broken blocks in the dam body. Large displacements and breakage zones in the backfill are created during the first filling. This is proof of large shear strains within the dam body which also affect the dry stone pitching. Indeed, we can note the dismantling of the berm which tends to increase in size. Nevertheless, a new state of stability is reached. This computation, performed by taking into account low characteristics for the rockfill, shows the beginning of the kinematics of the static mode of failure. The crest would move downstream, the upper part of the dry stone pitching would slide along the downstream surface of the rockfill dam body and would punch the berm and push the lower part of the dry stone pitching.

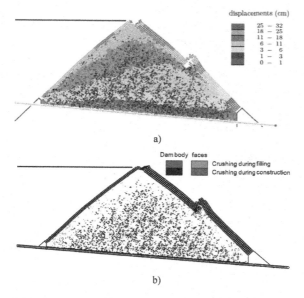

Figure 2.50. *Dam with block breakage after filling; a) isocontours of displacements and b) broken blocks with η=3.5.10⁵N. For a color version of the figure, see www.iste.co.uk/vincens/drystone.zip*

From these results, it can be concluded that:

– bulging of the dry stone pitching is caused by the compressibility of the dam body due to the breakage of blocks;

– even for very small breakage rates, settlement can be important;

– the unit strength $\eta=3.5*10^5$N leads to the dismantling of the dry stone pitching of the berm. This behavior is not observed at all on site;

– values of η higher than 5.10^5N are required to simulate a berm behavior in accordance with observed deformations on site.

2.4.2.4. Deformation analysis during 10 years of operation without aging

In the following modeling, the values for η are set to 10^6N (strength of a sounder granite) in a first computation and to 5.10^5 N (strength of a weaker granite) in a second one to simulate 10 years of operation without aging. These various strength configurations allow us to evaluate separately the effect of aging and compaction and hardening due to the cycles of reservoir loading on the dam behavior. The relative density is taken to $I_D=50\%$, value assumed representative of the density of the dam body. The friction coefficient between blocks μ is taken to 1, to obtain a mobilized friction angle at peak ϕ equal to 38°, under a confining stress of 100 kPa. Figures 2.51(a–b), respectively, show the percentage of breakage and the maximum displacements versus time on the downstream face at point 1 (crest) and at point 4 (berm).

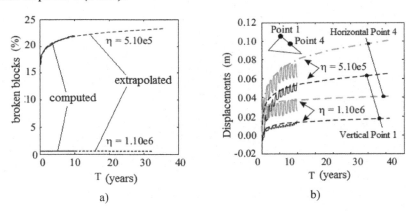

Figure 2.51. *Results of a 10 years of operation without aging; a) percentage of broken blocks; and b) computed displacements without aging. For a color version of the figure, see www.iste.co.uk/vincens/drystone.zip*

For the sound granite, after some rare breakages during the first filling and emptying, no more crushing is observed during the nine following years. In consequence, the displacements are more or less stabilized after 10 years of operation and settlements smaller than 0.1 mm/y will take place after 20 years.

It can be concluded from these results that without aging the sound rockfill does not present residual deformations. Residual deformations are specifically associated with aging, and not with the hardening and compaction of the rockfill under the cyclic loading.

On the contrary, for the computation involving the weaker granite, after the first year of operation, 17% of blocks are broken, and breakage goes on the following years with a logarithmic trend (0.7 mm/y after 20 years).

The maximum horizontal displacement is found to be twice the maximum vertical displacement. The elastic horizontal displacement is the same for both rockfills with about 1.5 cm. However, the elastic vertical displacement is equal to 5 mm for the weaker rockfill and 2 mm for the sound rockfill. The maximum elastic displacement is at the center of the upstream face.

2.4.2.5. Deformation analysis during 19 years of operation with aging

Three sets of parameters for the aging law are tested in this section. Aging-1 is based on the measured data published by [KRA 80] and aging-2 is based on data by [LAU 99]. They fit experimental data of strength reduction versus time up to several months and they represent the best estimate of short-term aging. Aging-1 leads to more breakage during dam construction and less breakage during dam operation than aging-2. Both aging models have the activation threshold F_{activ} equal to 40% of the strength at T=0. Aging-3 is not based on real data, but it is proposed as a very pessimistic case: all the blocks are broken at T=125 years. The aging parameters are given in Table 2.9. The time to failure of the blocks is reported for the three aging laws in Figure 2.52 versus the ratio α of applied stress:

$$\alpha = \frac{F_{fail}^{T=0} - F_{app}}{F_{fail}^{T=0} - F_{activ}}$$

[2.38]

Parameters	$\beta 1$	$\beta 2$	$\beta 3$	Tmax (years)
Aging-1	0	45	$3.1*10^{-14}$	$3.2*10^{+13}$
Aging-2	0	23	$3.1*10^{-7}$	$3.2*10^{+06}$
Aging-3	0	7	$8*10^{-3}$	125

Table 2.9. *Selected coefficients of the three aging laws*

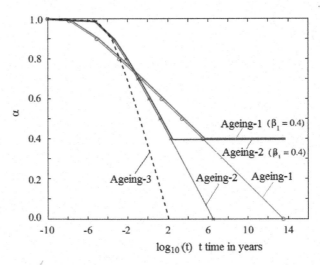

Figure 2.52. *Failure time versus the applied stress ratio α. For a color version of the figure, see www.iste.co.uk/vincens/drystone.zip*

Figures 2.53(a–d) show the impact of aging on the weakest rockfill: $\eta = 5.10^5$ N. The generation of the maximum horizontal displacements occurs at the downstream berm and is presented in Figure 2.53(a) and the maximum settlements occurring at crest are given in Figure 2.53(b). The ratio of the elastic deformation on the permanent deformation decreases with time: it means that the permanent deformations increase and hide the elastic behavior. At the fifth year of operation, the annual displacement increases from 4 cm/year up to 10 cm/year. This tertiary creep is the sign of the beginning of failure. Aging significantly increases the breakage of blocks (Figure 2.53(c)), after 10 years a third (aging-1 and 2) to a half of the total number of blocks are broken. For the aging-3 model, they become too large so that the rigid stone pitching cannot adapt them, and it breaks below the berm. In conclusion, when the initial strength of the rock is not sufficient, aging decreases this strength gradually in such a way that it causes permanent damage up to failure, whatever the law of aging.

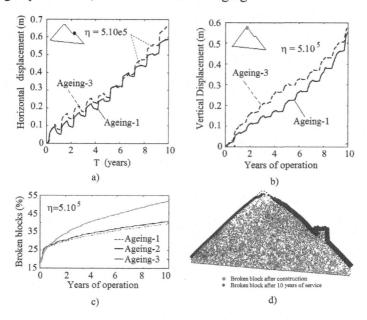

Figure 2.53. *Results of a 10 year operation with aging for η=5.105N: displacements a) at point 4; b) at point 1; c) percentage of breakage; d) repartition of breakage in the dam body. For a color version of the figure, see www.iste.co.uk/vincens/drystone.zip*

Figure 2.54 shows the impact of aging on the sound rockfill ($\eta = 10^6$). The generation of maximum horizontal displacements occurs at the downstream berm and is presented in Figure 2.54(a) and maximum settlements occurring at the crest are presented in Figure 2.54(b). The linear trend of the displacements versus time is noteworthy. The elastic deformations are quite constant with about 15 mm/year in the horizontal direction and about 2 mm in the vertical direction.

The set of parameters for the aging law derived from real data (aging-1 and aging-2) induces the development of very small permanent deformations in the dam with about 0.5 mm/year after 10 years of operation. These low deformations are associated with the very low amount of block breakage by aging (Figure 2.54(d)), far smaller than 1% (Figure 2.54(c)). The broken blocks are mainly at the base of the dam, where the stress levels are higher.

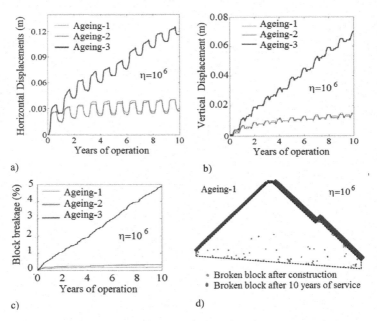

a)

b)

c)

d)

Figure 2.54. *Displacements and percentage of breakage after 10 years for η=106. Displacements a) at point 4; b) at point 1; c) percentage of breakage; d) repartition of breakage in the dam body. For a color version of the figure, see www.iste.co.uk/vincens/drystone.zip*

The set of parameters corresponding to the case of aging-3 (pessimistic case) leads to a rate of displacements which is constant with about 9 mm/year in the horizontal direction and 6 mm/year for vertical settlements. The origin of this trend must be found in the constant rate of the generation of broken blocks with time. In this case, 0.5% of the total number of blocks are broken per year. According to the assumption of the very severe set of parameters aging-3, all the blocks will be broken after 125 years of operation. This means that the linear trend will be followed by a tertiary creep, before all the blocks are broken.

In conclusion, even if the initial strength of the rocks is high, aging causes a permanent rate of displacement. However, in the case of parameters for the aging law derived from real data (aging-1 and aging-2), a very low permanent deformation rate is observed, lower than the current rate measured on actual dams. On the contrary, far larger deformation rates than measured on site are generated in the case of the pessimistic set of parameters for the aging law (aging-3). The real rate of deformations measured on actual dams is in fact between the two simulated rates.

2.4.2.6. *Deformation analysis during the dam lifetime*

According to the previous simulations of cyclic loading with aging, the unit strength $\eta=5.10^5$ N is no longer used. Higher strengths are selected with the set of parameters presented in Table 2.6: the first one for a normal or average rock and a best case representing a very sound rock. The simulations have been performed, in order to seize the effect of block breakage and aging separately, during a lifetime of 80 years [PLA 09b]. Five types of blocks are considered:

– unbreakable;

– normal rock;

– sound rock;

– normal rock weathered by aging effect;

– sound rock weathered by aging effect.

The vertical and horizontal displacements of the downstream crest of the dam are presented in Figures 2.55 and 2.56. Although the curves of displacement are scattered by some oscillations, three behaviors are observed:

– the permanent displacements with unbreakable blocks are very small, limited to 2 cm max. They are principally concentrated in the five first cycles. After this short period, the dam seems to be almost stabilized and the effect of the cyclic hardening is very limited;

– the permanent displacements of the rockfill made up of sound blocks are a little bit larger. Although weathering increases the permanent deformation to 4 cm in the horizontal direction and 6 cm in the vertical direction, the stabilization occurs with intact and weathered blocks;

– on the contrary, a linear trend for the evolution of deformation versus time is observed in the short term for the rockfill made up of normal blocks. However, for the very long term, the displacement rate increases with time.

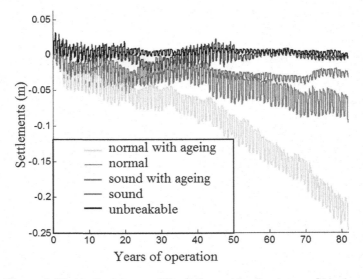

Figure 2.55. *Vertical creep of the D/S crest for five types of blocks. For a color version of the figure, see www.iste.co.uk/vincens/drystone.zip*

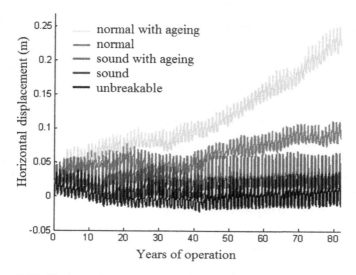

Figure 2.56. *Horizontal creep of the D/S crest for five types of blocks. For a color version of the figure, see www.iste.co.uk/vincens/drystone.zip*

The prediction of the model for the normal rock is very close to the monitored behavior (Table 2.10). The dam was built in 1953. The modern and reliable system of monitoring was installed in 1982. Thus, the monitored displacements can be compared to the computed displacements between 30 and 49 years of three benchmarks 19, 20 and 21 placed at the crest (Figure 2.18, dam B). The calculated displacements are in the range of the measured ones for the period 1983–2002. However, the extrapolation of the monitored data to the previous period 1953–1983 gives larger displacements than the model. If the model predicts an accurate rate of deformation in the long term, it underestimates the short-term behavior.

Period	Direction	Values		
		Monitored	Normal rock	Normal +aging
1953–1983	Horizontal (cm)	14?	5	9,5
	Settlements (cm)	11?	4	8
1983–2002	Horizontal (cm)	3.7–4.9	3,5	4
	Settlements (cm)	0.3–4.0	2,5	4

Table 2.10. *Comparison of monitored and computed displacements*

In conclusion, the aging of the rockfill causes a linear trend of deformation versus time. The model underestimates the short-term deformations, but it captures the long-term rate deformations quite well. In the very long term, for instance more than one century, the aging model predicts an increase in the rate of deformations.

2.4.2.7. Another scenario: model of the downstream face aging

The displacements calculated during the first 30 years of operation seem underestimated. A possible explanation could be an underestimation of the weathering. Aging is mainly caused by extreme changes of the climatic conditions (cycles of frost-thaw, dry and wet humidity [CHA 03b] and low and high temperatures). These severe conditions only occur at the downstream face of the dam, the crest and upstream facings are protected by an impervious concrete (Figure 2.57). Melnik [MEL 82] only found evidence of weathering of weak rockfill in the first 2–4 m. Thus, a more refined model involving a weathering of the rockfill in the first meters of the downstream face is proposed.

Two models of weathering by climatic impact are added to the aging process described in section 2.3.2.4:

– Climate-1: the dry stone pitching is weathered, and its friction coefficient μ is reduced from 1 to 0.7;

– Climate-2: the dry stone pitching and a 2.5 m thick layer parallel to the downstream slope and 1.5 m thick layer parallel to the upstream face of the random rockfill are weathered and obtain a reduced friction coefficient μ from 1 to 0.7 (Climate-2a). An extreme weathering was tested in Climate-2b, where all the friction coefficients were reduced to 0.7 (stone pitching and dam body).

The results show that the displacements are concentrated along the downstream face. It is interesting to note that the addition of surface weathering of Climate-1 and that of extreme weathering of Climate-2b to the aging-1 process increases by a factor 1.1 and 2.5, respectively, the maximum horizontal displacement and by a factor 1.1 and 2, respectively, the settlements occurring under cyclic loading with aging-1. These results suggest that the residual settlements could mainly occur in the dam body.

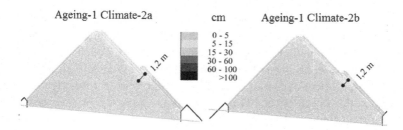

Figure 2.57. *Effect of surface weathering of the downstream face on global deformations. For a color version of the figure, see www.iste.co.uk/vincens/drystone.zip*

In conclusion, there is not a single model able to explain the displacements of the rockfill dams and several scenarios may reproduce the current rate of deformations. Other pieces of data from the dam are required to break down the causes and calibrate the parameters of the model.

2.4.3. *Seismic analysis*

Seismic simulations have been undertaken, because DEM seems particularly interesting to understand block displacements at the crest under large amplification [DEL 04, DEL 06]. Figure 2.54 shows that the largest displacements are located on the upstream facing of the dam. The accelerogram at the dam base is presented in Figure 2.58.

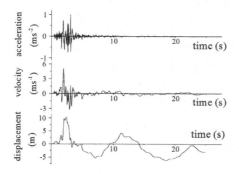

Figure 2.58. *Accelerogram, chronogram and displacement imposed at the dam base. For a color version of the figure, see www.iste.co.uk/vincens/drystone.zip*

The loading is applied as an imposed velocity to the basis of the dam. Here, the numerical damping coefficient has been reduced from 0.7 to 0.15. Indeed, the damping has to be reduced because of the dynamic nature of the sollicitation. However, it is still necessary to apply a small dissipation in order to dissipate high-frequency waves to stabilize the calculation. This value was empirically selected after testing its influence.

The properties of the rockfill are those presented in section 2.3.1.2. They are considered as a compromise between accuracy and computing time. Very high contact stiffnesses are required to transfer the dynamic waves propagating in the system: in a view of reducing the computing time they have been decreased, in such a way the elastic stiffnesses at contact are 10 times lower than the ones measured on site. It means that the model cannot catch the real vibratory behavior and underestimates the amplification in the elastic domain. However, the error decreases while the peak ground acceleration (PGA) increases, particularly in the strong motion phase, where the shear strains are higher than $>10^{-3}$. In this case, the deformations within the dam are mainly irreversible.

In a first computation, the reservoir is supposed to be empty. A sensitivity analysis was then carried out by changing the PGA (Figure 2.59). Three values are considered: 0.1, 0.3 and 0.5 g. The dynamic forces trigger block breakage and generate 7% of broken blocks under PGA = 0.5 g. The postseismic displacements are associated with the crushing, without crushing the postseismic displacements are 7 times less, for instance about 1.5 cm under PGA = 0.1 g. Without the hydrostatic pressure, the dynamic forces induce a bulging of the upstream face, the random rockfill settles and moves under the stone pitching. The crest and the berm are the two points of the downstream face where deformations are the highest. The berm is the weakest point of the downstream face, it might be punched under the seismic loads. The maximum displacement at crest is 9 cm with a PGA= 0.1 g, 16 cm with a PGA=0.3 g and 23 cm with a PGA=0.5 g. In all the situations, the dam is stable.

With a full reservoir, the horizontal deformations of the dam are larger, induced by larger stresses and in consequence by larger block breakage. There is a global downstream deformation of the dam body. The upstream face is confined by the hydrostatic pressure which prevents the formation of a bulging. It may explain why in this case the settlements are lower than when the reservoir is empty. Figure 2.60 shows the comparison of displacements at crest and at the middle part of the upstream facing.

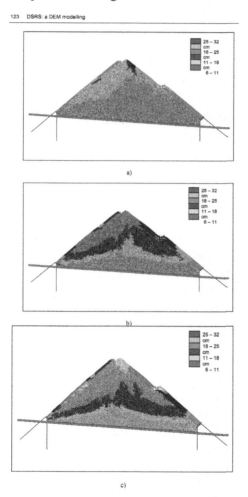

Figure 2.59. *Isocontours of post-seismic displacements; PGA: a) 0.1 g; b) 0.3 g and c) 0.5 g. For a color version of the figure, see www.iste.co.uk/vincens/drystone.zip*

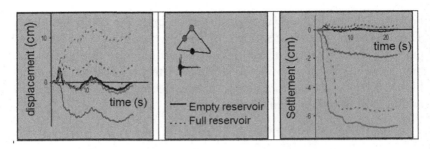

Figure 2.60. *Comparison of seismic displacements with an empty and a full reservoir. For a color version of the figure, see www.iste.co.uk/vincens/drystone.zip*

2.5. Physical tests for DEM model qualification

2.5.1. *Objectives*

The discrete elements model looks to be a powerful tool to capture the behavior of the rockfill. However, the validation of DEM procedures requires well-documented case studies. Up to now, it has not been possible to find in the dam literature a case study with all the required data for validation. With a view of getting a full set of assumed physical and mechanical properties, the calibration of parameters, involved in the DEM computation, requires several levels of justifications at three different scales: the scale of the block, triaxial test and structure performance.

A physical model of a granite rockfill dam with dry stone pitching at a 1/10 scale, loaded up to the failure, was planned within the research project PEDRA[1] and funded by EDF and supervised by a member of LTDS from *Ecole Nationale des Travaux Publics de l'Etat*. The properties of the rock, the mechanical properties of the triaxial tests and the performance of the dam are described hereafter to be a referenced case story used as a benchmark for a DEM model of a rockfill dam with a dry stone pitching.

1 PEDRA: behavior of dry stone and weakly bonded structures (2011–2015), Project C2D2, Ministry of Ecology, France.

2.5.2. *Physical properties of random rockfill*

The geology rock type is granite. The rock is sound. The rock was crushed, screened and washed.

The particle shapes are angular (Figure 2.61(a)). The grain size distribution is 10/80 mm with d_{50} equal to 40 mm, and Cu equal to 2 (Figure 2.61(b)).

a)

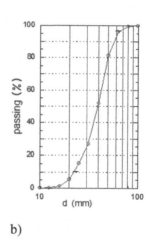

b)

Figure 2.61. *a) Photograph and b) grain size distribution of the rockfill. For a color version of the figure, see www.iste.co.uk/vincens/drystone.zip*

2.5.3. *Physical and mechanical properties of dry stone pitching*

The dry stone pitching is made up of cubic stone pavers with the dimensions $50 \times 50 \times 50$ mm, or rectangular stone pavers with the dimensions $50 \times 50 \times 100$ mm. The minimum and maximum widths of the stone pavers are, respectively, 40 and 60 mm.

The angle of friction between the stone pavers was measured by 10 repetitive tests between three pavers placed on three pavers. The average value for the contact friction angle Ψ between the pavers is found to be equal to 29° and the standard deviation is equal to 2°.

2.5.4. Mechanical properties of random rockfill

Three 1 m diameter and 1.5 m high triaxial samples of dumped rockfill were sheared at low density (Table 2.11) at laboratory GeM from Ecole Centrale de Nantes. The test PEDRA 3 can be considered as a repeatability test for the confinement 100 kPa. Strains larger than 10% are required to reach the strength of the samples. The friction angles are associated with the confining pressure and derived from Figure 2.62:

$\phi' = 42.3°$ at $\sigma'3 = 100$ kPa

$\phi' = 39.5°$ at $\sigma'3 = 200$ kPa

Test	Masse (kg)	Diameter (m)	Height (m)	γ_d (kg/m^3)	e_0	σ'_c (kPa)	$\varepsilon_{1,max}$ (%)
PEDRA 1	1662.4	0.993	1.504	14.00	0.86	100	12.5
PEDRA 2	1680.3	0.989	1.502	14.29	0.82	200	17.3
PEDRA 3	1660.0	0.987	1.503	14.16	0.84	100	17.3

Table 2.11. *Initial properties for the samples and conditions for the triaxial tests*

2.5.5. Angle of repose of the slope of the random rockfill

A first test was carried out to find the angle of repose of the random rockfill. Four deformation cells were glued on stones of the downstream face. The rockfill was dumped in a tipper truck with an initial downstream slope of 24° and height of 1.7 m. The tipper is 5.6 m long, 2.25 m wide and 2.1 m high and was rotated by a step of 1°. The step lasts long enough to allow stabilization.

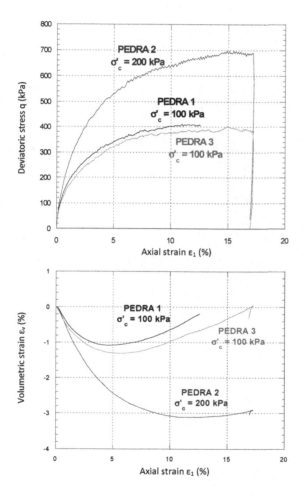

Figure 2.62. *Triaxial curves on the rockfill versus vertical strain (%); a) deviatoric stress; b) volumetric strains. For a color version of the figure, see www.iste.co.uk/vincens/drystone.zip*

It is impressive to observe the creep increasing with the decrease in the margin of safety (Figure 2.63). This creep was noticed by the DEM modeling. The physical model confirms the numerical model. For both models, the question that arises is, is the real value of the angle of repose following a very long-term creep? From 24 to 36°, no deformation is observed. Deformation of the slope begins as soon as the angle of inclination of the tipper passes 12°, i.e. slope angle $\phi=24+12=36°$. From 36 to 41°, the displacements of the slope are simultaneous to the rotation of the tipper. From 41 to 46°, the deformations are mainly observed during the step time where the tipper does no rotate. After 46°, the slope slides.

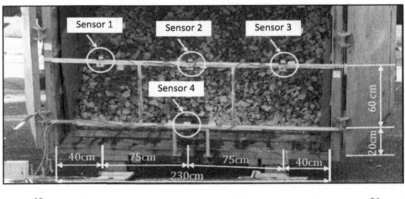

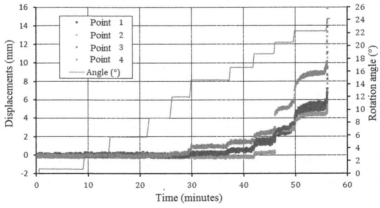

Figure 2.63. *Cells and rotation and displacements of the rockfill dam slope. For a color version of the figure, see www.iste.co.uk/vincens/drystone.zip*

2.5.6. *Angle of repose of the rockfill dam with dry stone pitching*

The rockfill dam is built on the same tipper truck as previously mentioned. The dam is 2 m high, with a slope H/V=1/1 and a 0.2 m wide crest (Figure 2.64). It is covered on its upstream and downstream faces by one 5 cm thick layer of dry stone pitching (Figure 2.65). Six benchmarks are installed on the downstream face to measure the deformation of the dry stone pitching.

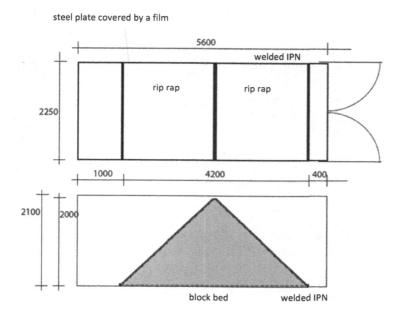

Figure 2.64. *Measured rotation and displacements of the rockfill slope. For a color version of the figure see www.iste.co.uk/vincens/drystone.zip*

The tipper is rotated one degree by one degree at the beginning and by half a degree after the first deformations have been observed. The permanent deformation is the sum of the deformation at the end of the rotation (instantaneous deformation) and the creep deformation (at constant condition). During the failure tilting test (Figure 2.66), the main observations are:

– the downstream face begins to deform at $\alpha = 45°+16° = 57°$;

– creep deformations are larger than deformations at the end of rotation at $\alpha = 45°+19° = 64°$;

– collapse is triggered at $\alpha = 45°+24° = 69°$;

– the dry stone pitching is an efficient facing that to a large extent increases the safety coefficient from FS = 1.03 to FS = 2.52 (tan(69)/tan(46)).

This failure test confirms the valuable role of the dry stone pitching confining and reinforcing the downstream face of the rockfill.

Figure 2.65. *Construction of the dry stone pitching on the rockfill dam, For a color version of the figure, see www.iste.co.uk/vincens/drystone.zip*

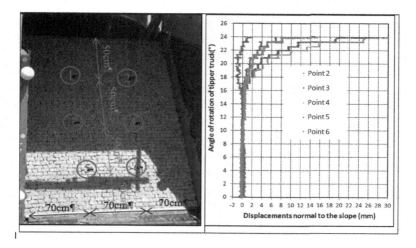

Figure 2.66. *Measured rotation and displacements of the physical model of dam. For a color version of the figure, see www.iste.co.uk/vincens/drystone.zip*

2.6. Conclusion

The CFRDs built in France in the first part of the 20th Century are quite different from the current normalized design standards. They have very steep slopes and their upstream and downstream faces are confined with a thin dry stone pitching. In consequence, it is not surprising to notice quite a different behavior from the other CFRDs. First, from place to place breakages of stones are noticed. When stone breakages are concentrated in a given area, they can lead to the pitching dismantling with extrusion of one or several stones. Sometimes, rows of buckling and bulging appear in the downstream stone pitching of some dams. Finally, some linear deformations versus time are monitored and ask for the question of the risk of failure. To understand this behavior, first a compilation of rockfill and dry stone pitching properties were encompassed. The scale and time effects appear to be the most important phenomena, underestimated by the current methods of stability analysis.

To change this situation, a DEM was proposed. It was implemented in a commercial software (PFC^{2D}, code ITASCA). The first test was done in 1997: it demonstrated that triaxial test curves of rockfill can be successfully simulated with the DEM. Thus, the main point was to catch the physical meaning of the particle breakage: to simulate (1) the increase in particle strength and decrease in block failure stress with the block size, (2) the delayed fracture or decrease in particle strength with time. Two major difficulties were encountered by the modeling: (1) the breakage and triaxial tests measure 3D phenomena and PFC^{2D} analyzes 2D phenomena, thus a physically based 3D/2D transfer rule has to be adopted; (2) the dam instability follows a transient creep period and the determination of the critical loading requires numerical thresholds. The calibration requires three stages at three different scales: the particle scale (for the calibration of particles static failures in Marsal breakage tests), the medium scale (for the calibration of the friction and the stiffness coefficients in Marsal triaxial tests) and the dam scale (where both previous calibrations are mixed and implemented). This approach was applied on an existing dam by simulating two rockfill behaviors: one being a weak rockfill and the other one being a very sound rockfill. The influence of climate change was included in the aging process of the dry stone pitching.

The comparison of the dam displacements with the modeled ones, after the first filling, permitted a satisfying calibration. The comparison of creep displacements after 30 years was quite good. So, the application of the model to the dam operation for one century leads to the following conclusions: (1) with the very sound rockfill, the compaction under cyclic reservoir loadings without block breakage leads to negligible displacements, (2) with the weak rockfill, the coupling of cyclic reservoir loadings and the breakage of blocks leads to a constant displacement rate followed by a rate increasing with time, showing that local or global repair of the dry stone pitching will be required in the long term. The modeling of a seismic loading shows acceptable displacements up to large PGA. In conclusion, the DEM is very useful for the study of the behavior of such a dam. Applications for other dams and other loadings are waiting for final static and seismic calibration tests. In this sense, the first physical model of dam failure at scale 1/10 and its subsequent modeling that must be done in the future will be very helpful to check the suitability and validity of the whole numerical process (physical assumptions and numerical procedures) to model other CFRDs with the DEM.

Conclusion

The discrete element method (DEM) modeling of complex structures like the one involved in this book cannot be addressed without access to a vast amount of data related to the behavior of the rocky materials, as well as to the permanent deformation of the actual structures monitored during their lifetime. Nevertheless, some key information on site including the state of alteration of the rockfill material for the case of rockfill dams or the evolution of the conditions at the boundaries (earth pressure and deflection of the foundation) in the case of DSRWs is difficult to obtain.

The development of down-scale experiments ultimately became critical to step up the knowledge of the scientific community on the behavior of such structures. Very specific skills were required to develop and the complexity of the experiments has not always made it easy to get either a quick or definite answer to all the questions raised by the observations on actual structures. Further research is required in this respect. Since down-scale experiments often require the simplification of the characteristics or properties of the actual systems to be processed, challenges still have to be met to obtain a clear set of data for the validation of DEM models using full-scale experiments.

DEM has allowed us to get information that other methods would not provide or would provide with poor accuracy. We can cite the deformation field for the highway DSRWs, the selective breakage of blocks and the bulging of the pitching in rockfill dams. Nevertheless,

DEM being a sophisticated numerical tool, modeling such complex systems involves a very time-consuming calibration of model parameters. Moreover, computation time for studying the mechanical behavior of these systems can be huge, which is specifically the case of rockfill dams due to the elevated number of degrees of freedom and the possibility for the rockfill to crush. Thus, a fully DEM approach may not be adapted in all the cases and the modeling of some relevant parts of the system using a finite element method can partly answer the issue.

At this stage, for DSRWs, the short-term challenges to bet met are related to the design of highway DSRWs even if some partial results are available. Another critical aspect is related to the stability of slope DSRWs when heavy water flows are involved. The effect of aging of the rocky material is an issue that is shared with rockfill dams with dry stone pitching. Aging alters contact points and creates further degrees of freedom.

Bibliography

[ALE 12] ALEJANO L.R., VEIGA M., TABOADA J. et al., "Stability of granite drystone masonry retaining walls: I. Analytical design", *Géotechnique*, vol. 62, no. 11, pp. 1013–1025, 2012.

[AUV 75] AUVINET G.G., MARSAL R.J., "Statistical model of grain breakage", *5th Pan-American Conference on Soil Mechanics and Foundation Engineering*, Buenos Aires, vol. 1, pp. 193–204, 1975.

[AZÉ 10] AZÉMA E., RADJAÏ F., "Stress-strain behavior and geometrical properties of packings of elongated particles", *Physical Review E*, vol. 81, p. 051304, 2010.

[BAR 81] BARTON N., KJAERNSLI B., "Shear strength of rockfill", *Journal of the Geotechnical Engineering Division*, ASCE, vol. 107, no. 7, pp. 873–891, 1981.

[BEL 09] BELHEINE N., PLASSIARD J.P., DONZÉ F.V. et al., "Numerical simulation of drained triaxial test using 3D discrete element modeling", *Computers and Geotechnics*, vol. 36, no. 1, pp. 320–331, 2009.

[BIA 94] BIAREZ J., HICHER P.-Y., *Elementary Mechanics of Soil Behaviour: Saturated Remoulded Soils*, AA Balkema, 1994.

[BIA 97] BIAREZ J., HICHER P.Y., "Influence de la granulométrie et de son évolution par ruptures de grains sur le comportement mécanique des matériaux granulaires", *Revue Française de Génie Civil*, vol. 1, no. 4, pp. 607–631, 1997.

[BOL 86] BOLTON M.D., "The strength and dilatancy of sands", *Géotechnique*, vol. 36, no. 1, pp. 65–78, 1986.

[BON 15] BONILLA-SIERRA V., SCHOLTÈS L., DONZÉ F.V. *et al.*, "Rock slope stability analysis using photogrammetric data and DFN–DEM modelling", *Acta Geotechnica*, vol. 10, no. 4, pp. 497–511, 2015.

[CAL 03] CALVETTI F., VIGGIANI G., TAMAGNINI C., "A numerical investigation of the incremental non-linearity of granular soils", *Italian Geotechnical Journal, Special Issue on Mechanics and Physics of Granular Materials*, vol. 3, no. 3, pp. 11–29, 2003.

[CHA 58] CHARLES R.J., "Static fatigue of glass", *J. of Applied Physics*, vol. 29, 1549, 1958.

[CHA 03a] CHAREYRE B., Modélisation du comportement d'ouvrages composites sol-géosynthétique par éléments discrets: application aux ancrages en tranchées en tête de talus, PhD Thesis, University Joseph-Fourier-Grenoble I, 2003.

[CHA 03b] CHAVEZ C., ALONSO E.E., "A constitutive model for crushed granular aggregates which includes suction effects", *Soil and Foundation*, vol. 43, no. 4, pp. 215–227, 2003.

[CLA 05] CLAXTON M., HART R.A., MCCOMBIE P.F. *et al.*, "Rigid block distinct-element modeling of dry-stone retaining walls in plane strain", *Journal of Geotechnical and Geoenvironmental Engineering*, vol. 131, no. 3, pp. 381–389, 2005.

[CLE 84] CLEMENT R.P., "Post-construction deformation of rockfill dams", *Journal of Geotechnical Engineering*, vol. 110, no. 7, pp. 821–840, 1984.

[COL 10a] COLAS A.-S., MOREL J.-C., GARNIER D., "Full-scale field trials to assess dry-stone retaining wall stability", *Engineering Structures*, vol. 32, no. 5, pp. 1215–1222, 2010.

[COL 10b] COLAS A.-S., MOREL J.-C., GARNIER D., "2D modeling of a dry joint masonry wall retaining a pulverulent backfill", *International Journal for Numerical and Analytical Methods in Geomechanics*, vol. 34, no. 12, pp. 1237–1249, 2010.

[COY 39] COYNE A., *Leçons sur les Grands Barrages*, Ecole Nationale des Ponts et Chaussées, 1939.

[CUN 79] CUNDALL P.A., STRACK O.D.L., "A discrete numerical model for granular assemblies", *Géotechnique*, vol. 29, no. 1, pp. 47–65, 1979.

[DAN 02] DANO C., HICHER P.-Y., "Evolution of elastic shear modulus in granular materials along isotropic and deviatoric stress path", *Proceedings of the 15th ASCE Engineering Mechanics Conference*, New York, NY, pp. 1–8, 2–5 June 2002.

[DEL 02] DELUZARCHE R., CAMBOU B., FRY J.-J., "Modelling of rock-fill behaviour with crushable particles", *1st International PFC Symposium Gelsenkirchen Germany*, Balkema, 2002.

[DEL 04] DELUZARCHE R., Modélisation discrète des enrochements. Application aux barrages, PhD Thesis, Ecole Centrale de Lyon, 2004.

[DEL 06] DELUZARCHE R., CAMBOU B., "Discrete numerical modelling of rockfill dams", *International Journal for Numerical and Analytical Methods in Geomechanics*, vol. 30, no. 11, pp. 1075–1096, 2006.

[DIC 96] DICKENS J.G., WALKER P.J., "Use of distinct element model to simulate behaviour of dry-stone wall", *Structural Engineering Review*, vol. 8, nos. 2–3, pp. 187–199, 1996.

[FIO 02] FIORAVANTE V., "On the shaft friction modelling of non-displacement piles in sands", *Soils and Foundations*, vol. 42, no. 2, pp. 23–33, 2002.

[FLA 90] FLAVIGNY E., DESRUES J., PALAYER B., "Le sable d'Hostun RF", Note technique, *Revue française de géotechnique*, vol. 53, pp. 67–70, 1990.

[FRO 05] FROSSARD A., "Comportement macroscopique des matériaux granulaires mis dans les barrages", *XVIIth Projet de recherché Microbe*, p. 75, 30 May 2005.

[FRO 09] FROSSARD A., "Scale effects in granular fill shear strength and in stability of large rockfill structures", *17th International Conference on Soil Mechanics and Geotechnical Engineering*, Alexandria, Egypt, 5–9 October 2009.

[GAR 85] GARBRECHT G., "Sadd-El-Kaffara, the world oldest large dam", *International Water Power and Dam Construction*, pp. 71–76, July 1985.

[HAR 85] HARDIN B.O., "Crushing of soil particles", *Journal of the Geotechnical Engineering Division*, ASCE, vol. 111, no. 10, pp. 1177–1192, 1985.

[HAR 00] HARKNESS R.M., POWRIE W., ZHANG X. *et al.*, "Numerical modelling of full-scale tests on drystone masonry retaining walls", *Géotechnique*, vol. 50, no. 2, pp. 165–179, 2000.

[HAR 12] HARTHONG B., SCHOLTÈS L., DONZÉ F.-V., "Strength characterization of rock masses, using a coupled DEM-DFN model", *Geophysical Journal International*, vol. 191, no. 2, pp. 467–480, 2012.

[ICO 10] ICOLD, CIGB, "Concrete Face Rockfill Dam: Concepts for Design and Construction", *Bulletin 141*, Paris, CIGB, 2010.

[ITA 06] ITASCA CONSULTING GROUP, INC., UDEC (Universal Distinct Element Code), Version 4.0., Minneapolis, ICG, 2006.

[ITA 08] ITASCA CONSULTING GROUP INC., *Particle Flow Code in 2 Dimensions: Theory and Background*, 4th ed., Itasca, Minnesota, 2008.

[IWA 98] IWASHITA K., ODA M., "Rolling resistance at contacts in simulation of shear band development by DEM", *Journal of Engineering Mechanics*, vol. 124, pp. 285–292, 1998.

[KIS 87] KISHIDA H., UESUGI M., "Tests of the interface between sand and steel in the simple shear apparatus", *Géotechnique*, vol. 37, no. 1, pp. 45–52, 1987.

[KRA 80] KRANZ R.L., "Effects of confining presssure and stress difference on static fatigue of granite", *Journal of Geophysical Research*, vol. 85, no. B4, pp. 1854–1866, 1980.

[LAD 10] LADE P.V., KARIMPOUR H., "Static fatigue controls particle crushing and time effects in granular materials", *Soil and Foundation*, vol. 50, no. 5, pp. 573–583, 2010.

[LAU 97] LAU J.S.O., CONLON B., GORSKI B., Long-term loading tests on saturated Lac Du bonnet pink granite, Note, AECL, Researchs, MMSL, vol. 76, 1997.

[LE 13] LE H.H., Stabilité des murs de soutènement routiers en pierre sèche: Modélisation 3D par le calcul à la rupture et expérimentation échelle 1, PhD Thesis, ENTPE, 2013.

[LEM 90] LEMAITRE J., CHABOCHE J.L., *Mechanics of Solid Materials*, Cambridge University Press, Cambridge, 1990.

[LEP 70] LEPS T.M., "Review of shearing strength of rockfill", *Journal of Soil Mechanics and Foundation Division*, ASCE, vol. 96, no. 4, pp. 1159–1170, 1970.

[LIU 13] LIU Z., KOYI H.A., "Kinematics and internal deformation of granular slopes: insights from discrete element modeling", *Landslides*, vol. 10, no. 2, pp. 139–160, 2013.

[LOR 09] LORIG L.J., WATSON A.D., MARTIN C.D. *et al.*, "Rockslide run-out prediction from distinct element analysis", *Geomechanics and Geoengineering*, vol. 4, no. 1, pp. 17–25, 2009.

[MAH 96] MAHBOUBI A., GHAOUTI A., CAMBOU B., "La simulation numérique discrète du comportement des matériaux granulaires", *Revue française de géotechnique*, no. 76, pp. 45–61, 1996.

[MAH 97] MAHBOUBI A., FRY J.-J., CAMBOU B., "Numerical modeling of the mechanical behavior of non spherical, crushable particles", *Powder and Grains*, pp. 139–144, 1997.

[MAM 89] MAMBA M., Résistance au cisaillement des enrochements et des matériaux grossiers: application aux calculs des barrages, PhD Thesis, Université des Sciences et Techniques de Lille Flandres Artois, 1989.

[MAR 67] MARSAL R.J., "Large-scale testing of rockfill materials", *Journal of Soil Mechanics and Foundation Division*, ASCE, vol. 93, no. 2, pp. 27–44, 1967.

[MAR 73] MARSAL R.J., "Mechanical properties of rockfill", in HIRSCHFELD R.C., POULOS S.J. (eds), *Embankment Dam Engineering: Casagrande Volume*, Wiley, New York, pp. 109–200, 1973.

[MAS 08] MAS IVARS D., POTYONDY D., PIERCE M. *et al.*, "The smooth-joint contact model", *Proceedings of the 8th World Congress on Computational Mechanics and 5th European Congress on Computational Methods in Applied Sciences and Engineering*, Venice, Italy, Paper No. a2735, 2008.

[MCD 98] MCDOWELL G.R., BOLTON M.R., "On the micromechanics of crushable aggregates", *Geotechnique*, vol. 48, no. 5, pp. 667–679, 1998.

[MEL 82] MELNIK V.G., "Use of low-strength soils for constructing dams", *Gidrotekhnicheskoe Stroitielstvo*, no. 8, pp. 31–37, 1982.

[MIR 02] MIRGHASEMI A.A., ROTHENBURG L., MATYAS E.L., "Influence of particle shape on engineering properties of assemblies of two-dimensional polygon-shaped particles", *Géotechnique*, vol. 52, no. 3, pp. 209–217, 2002.

[MOA 08] MOAREFVAND P., VERDEL T., "The probabilistic distinct element method", *International Journal for Numerical and Analytical Methods in Geomechanics*, vol. 32, no. 5, pp. 559–577, 2008.

[MOL 12] MOLLON G., RICHEFEU V., VILLARD P. *et al.*, "Numerical simulation of rock avalanches: influence of a local dissipative contact model on the collective behavior of granular flows", *Journal of Geophysical Research: Earth Surface*, vol. 117, no. 2, art. no. F02036, 2012.

[MOR 94] MOREAU J.J., "Some numerical methods in multibody dynamics: application to granular materials", *Eur. J. Mech. A (Solids)*, vol. 13, no. 4, pp. 93–114, 1994.

[MOR 07] MORTARA G., MANGIOLA A., GHIONNA V.N., "Cyclic shear stress degradation and post-cyclic behaviour from sand-steel interface direct shear tests", *Canadian Geotechnical Journal*, vol. 44, no. 7, pp. 739–752, 2007.

[MUN 10] MUNDELL C., McCOMBIE P., HEATH A. *et al.*, "Behaviour of drystone retaining structures", *Proceedings of the ICE – Structures and Buildings*, vol. 163, no. 1, pp. 3–12, 2010.

[NIC 07] NICOT F., GOTTELAND P., BERTRAND D. *et al.*, "Multiscale approach to geo-composite cellular structures subjected to rock impacts", *International Journal for Numerical and Analytical Methods in Geomechanics*, vol. 31, no. 13, pp. 1477–1515, 2007.

[NOO 14] NOORIAN BIDGOLI M., JING L., "Anisotropy of strength and deformability of fractured rocks", *Journal of Rock Mechanics and Geotechnical Engineering*, vol. 6, no. 2, pp. 156–164, 2014.

[NOU 03] NOUGUIER-LEHON C., VINCENS E., CAMBOU B., "Influence of particle shape and angularity on the behaviour of granular materials: a numerical analysis", *International Journal for Numerical and Analytical Methods in Geomechanics*, vol. 27, pp. 1207–1226, 2003.

[NOU 05] NOUGUIER-LEHON C., VINCENS E., CAMBOU B., "Structural changes in granular materials: the case of irregular polygonal particles", *International Journal of Solids and Structures*, vol. 42, nos. 24–25, pp. 6356–6375, 2005.

[NOU 10] NOUGUIER-LEHON C., "Effect of the grain elongation on the behaviour of granular materials in biaxial compression", *Comptes Rendus Mécanique*, vol. 38, no. 10–11, pp. 587–595, 2010.

[ODE 00] ODENT N., "Recensement des ouvrages de soutènement en bordure du réseau routier national", *Ouvrages d'art – centre des techniques d'ouvrages d'art*, vol. 34, pp. 15–18, 2000.

[OET 14] OETOMO J.J., Comportement à la rupture des murs de soutènement en pierre sèche: une modélisation discrète, PhD Thesis, Ecole centrale de Lyon, 2014.

[OET 15] OETOMO J.J., VINCENS E., DEDECKER F. *et al.*, "Modeling the 2D behavior of drystone retaining walls by a fully discrete element method", *International Journal for Numerical and Analytical Methods in Geomechanics*, 2015.

[ORE 99] O'REILLY M.P., BUSH D.I., BRADY K.C. *et al.*, "The stability of drystone retaining walls on highways", *Proceedings of the ICE – Municipal Engineer*, vol. 133, no. 2, pp. 101–107, 1999.

[PLA 10] PLASSIARD J.P., DONZÉ F.V., "Optimizing the design of rockfill embankments with a discrete element method", *Engineering Structures*, vol. 32, no. 11, pp. 3817–3826, 2010.

[PLA 09a] PLASSIARD J.-P., DONZÉ F.V., "Rockfall impact parameters on embankments: a discrete element methods analysis", *Structural Engineering International*, vol. 19, no. 3, pp. 333–341, 2009.

[PLA 09b] PLASSIARD J.P., FRY J.J., DEDECKER F., "Bidimensional discrete element simulations of ageing effects in rockfill dams", *LTBD09*, Graz 2009.

[POW 02] POWRIE W., HARKNESS R.M., ZHANG R.M. *et al.*, "Deformation and failure modes of drystone retaining walls", *Géotechnique*, vol. 52, no. 6, pp. 435–446, 2002.

[PRA 13] PRA-AI S., Behaviour of soil-structure interfaces subjected to a large number of cycles: application to piles, PhD Thesis, University of Grenoble, 2013.

[QUE 15] QUEZADA J.-C., VINCENS E., MOUTERDE R. *et al.*, "3D failure of a scale down dry stone retaining wall: a DEM modeling", *Engineering Structures*, 2015.

[ROT 92] ROTHENBURG L., BATHURST R.J., "Micromechanical features of granular assemblies with planar elliptical particles", *Géotechnique*, vol. 42, no. 1, pp. 79–95, 1992.

[RUS 60] RÜSCH H., "Researches toward a general flexural theory for structural concrete", *J. Am. Concr. Inst.*, vol. 57, no. 1, pp. 1–28, 1960.

[SAL 09] SALOT C., GOTTELAND P., VILLARD P., "Influence of relative density on granular materials behavior: DEM simulations of triaxial tests", *Granular Matter*, vol. 11, no. 4, pp. 221–236, 2009.

[SAL 10] SALCIARINI D., TAMAGNINI C., CONVERSINI P., "Discrete element modeling of debris-avalanche impact on earthfill barriers", *Physics and Chemistry of the Earth*, vol. 35, nos. 3–5, pp. 172–181, 2010.

[SCH 78] SCHMIDTKE R.H., LAJTAI E.Z., "Constitution of mechanics of granular materials through the graph theory", *US – Japan Seminar on Continuum – Mechanics and Statistical Approaches in the Mechanics of Granular materials*, pp. 47–62, 1978.

[SCH 94] SCHNITTER N.J., *History of Dams: The Useful Pyramids*, AA Balkema, 1994.

[SIL 09] SILVANI C., DÉSOYER T., BONELLI S., "Discrete modelling of time-dependent rockfill behavior", *International Journal for Numerical and Analytical Methods in Geomechanics*, vol. 33, no. 5, pp. 665–685, 2009.

[TAN 09] TANG C.L., HU J.C., LIN M.L. *et al.*, "The Tsaoling landslide triggered by the Chi-Chi earthquake, Taiwan: insights from a discrete element simulation", *Engineering Geology*, vol. 106, nos. 1–2, pp. 1–19, 2009.

[THO 00] THORNTON C., "Numerical simulations of deviatoric shear deformation of granular media", *Géotechnique*, vol. 50, no. 1, pp. 43–53, 2000.

[TIN 95] TING J.M., MEACHUM L., ROWELL J.D., "Effect of particle shape on the strength and deformation mechanisms of ellipse-shaped granular assemblages", *Engineering Computations*, vol. 12, no. 2, pp. 99–108, 1995.

[TRA 06] TRAN T.H., Analyse et modélisation du vieillissement des barrages en enrochement par une approche micromécanique, PhD Thesis, Ecole centrale de Lyon, 2006.

[TRA 09] TRAN T.-H., VENIER R., CAMBOU B., "Discrete numerical modelling of rock-ageing in rockfill dams", *Computers and Geotechnics*, vol. 36, nos. 1–2, pp. 264–275, 2009.

[UES 86] UESUGI M., KISHIDA H., "Influential factors of friction between steel and dry-sands", *Soils and Foundations*, vol. 26, no. 2, pp. 33–46, 1986.

[VIL 07] VILLEMUS B., MOREL J.-C., BOUTIN C., "Experimental assessment of dry stone retaining wall stability on a rigid foundation", *Engineering Structures*, vol. 29, no. 9, pp. 2124–2132, 2007.

[WAL 07] WALKER P., MCCOMBIE P., CLAXTON M., "Plane strain numerical model for drystone retaining walls", *Proceedings of the ICE – Geotechnical Engineering*, vol. 160, no. 2, pp. 97–103, 2007.

[WEI 51] WEIBULL W., "A statistical distribution function of wide applicability", *Journal of Applied Mechanics*, no. 18, pp. 293–297, 1951.

[YAN 12] YANG Z.X., YANG J., WANG L.Z., "On the influence of inter-particle friction and dilatancy in granular materials: a numerical analysis", *Granular Matter*, vol. 14, no. 3, pp. 433–447, 2012.

[ZHA 07] ZHANG L., THORNTON C., "A numerical examination of the direct shear test", *Géotechnique*, vol. 57, no. 4, pp. 343–354, 2007.

[ZHA 11] ZHAO Z., JING L., NERETNIEKS I. *et al.*, "Numerical modeling of stress effects on solute transport in fractured rocks", *Computers and Geotechnics*, vol. 38, no. 2, pp. 113–126, 2011.

Index

Printed in the United States
By Bookmasters